BUILDER−TESTED | CODE APPROVED

FRAMING FLOORS, WALLS & CEILINGS

FROM THE EDITORS OF Fine Homebuilding

The Taunton Press

The Taunton Press
Inspiration for hands-on living®

The Taunton Press, Inc., 63 South Main Street, PO Box 5506, Newtown, CT 06470-5506
e-mail: tp@taunton.com

Editor: Jennifer Renjilian Morris
Copy editor: Seth Reichgott
Indexer: James Curtis
Cover design: Alexander Isley, Inc.
Interior design: Cathy Cassidy
Layout: Kimberly Shake
Front cover photographer: Roe A. Osborn
Back cover photographer: Fernando Pagés Ruiz

Taunton's For Pros By Pros® is a trademark of The Taunton Press, Inc., registered in the U.S. Patent and Trademark Office.

Library of Congress Cataloging-in-Publication Data
Framing floors, walls & ceilings : from the editors of Fine homebuilding.
      p. cm. -- (Taunton's for pros by pros)
   Includes index.
   ISBN 978-1-60085-069-1
   1. House framing. I. Fine homebuilding.
   TH2301.F744 2008
   694'.2--dc22

                        2008036276

Printed in the United States of America
10 9 8 7 6 5 4 3 2 1

Except for new page numbers that reflect the organization of this collection, these articles appear just as they did when they were originally published. You may find that some information about manufacturers or products is no longer up to date. Similarly, building methods change and building codes vary by region and are constantly evolving, so please check with your local building department.

Construction is inherently dangerous. Using hand or power tools improperly or ignoring safety practices can lead to permanent injury or even death. Don't try to perform operations you learn about here (or elsewhere) unless you're certain they are safe for you. If something about an operation doesn't feel right, don't do it. Look for another way. We want you to enjoy the craft, so please keep safety foremost in your mind whenever you're in the shop.

Special thanks to the authors, editors, art directors, copy editors, and other staff members of *Fine Homebuilding* magazine who contributed to the development of the articles in this book.

CONTENTS

PART 2: MATERIALS

PART 3: FRAMING FLOORS

PART 4: FRAMING WALLS

PART 5: FRAMING CEILINGS

INTRODUCTION

According to John Lienhard, the host of public radio's *Engines of Our Ingenuity*, two things made building in Colonial America different from building in Europe. One was an abundance of wood. The other was a lack of skilled labor. To settle this country, we needed to invent a new way of building that required less skill and took advantage of local resources. We eventually did it, but two things had to happen first.

In 1791, Samuel Briggs patented a machine for making nails, and in 1813, a Shaker named Tabitha Babbitt invented the first circular saw used on a saw mill. These inventions set the stage for Augustine Taylor, in 1833, to build the first structure in America using what came to be known as balloon framing—small-dimension lumber held together with nails. Almost 200 years later we're still using that same method to frame houses.

Some things, of course, have changed. Balloon framing, eventually gave way to platform framing. Engineered materials were developed to account for the fact that wood is not nearly as abundant as it once was. And power tools make assembling easier and faster than ever before.

Unfortunately, something else has changed over the years: Interest in the building trades has declined. And we are once again suffering a lack of skilled labor, just like our Colonial ancestors. As a result, good information about building is hard to find, which is what makes this book so valuable. Collected here are 32 articles from past issues of *Fine Homebuilding* magazine. Written by professional builders from all over the country, these articles deliver detailed advice about framing floors, walls, and ceilings.

Kevin Ireton, editor,
Fine Homebuilding

10 Rules for Framing

■ BY LARRY HAUN

It was a coincidence that another contractor and I began framing houses next door to one another on the same day. But by the time his house was framed, mine was shingled, wired, and plumbed. It was no coincidence that the other contractor ran out of money and had to turn the unfinished house over to the lending company, whereas I sold mine for a profit.

Both houses were structurally sound, plumb, level, and square, but every 2×4 in the other house was cut to perfection. Every joint looked like finish carpentry. The other contractor was building furniture, and I was framing a house.

Unlike finish carpentry, framing doesn't have to look perfect or satisfy your desire to fit together two pieces of wood precisely. Whether you're building a house, an addition, or a simple wall, the goals when framing are strength, efficiency, and accuracy. Following the building codes and the blueprints should take care of the strength; efficiency and accuracy are trickier. But during 50 years of framing houses, I've come up with the following rules to help me do good work quickly and with a minimum of effort.

1. Don't Move Materials Any More Than You Have To

Hauling lumber from place to place is time-consuming and hard on your body. Make it easier on yourself every chance you get, and start by having the folks at the lumberyard do their part. Make sure lumber arrives on the truck stacked in the order it will be used (see the top illustration on the facing page). You don't want to move hundreds of wall studs to get to your plate stock, for instance. And floor joists go on top of floor sheathing, not the other way around.

When it's time for the delivery, unload the building materials as close as possible to where they will be used. Often lumber can be delivered on a boom truck, so stacks of lumber can be placed right up on the deck or on a simple structure built flush alongside the deck.

Once the material is delivered, don't move it any more than you need to. Cut studs, plywood, and anything else you can right on the stack. If you do have to move wood, plan so that you have to move it only once (see the bottom illustration on the facing page).

Loading Lumber

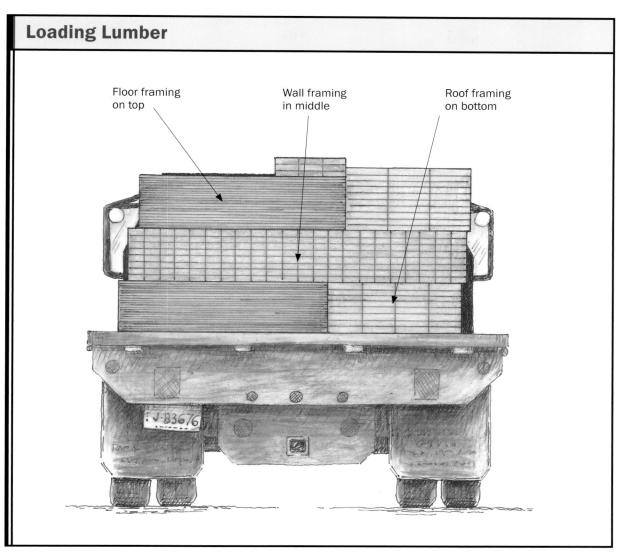

Floor framing on top

Wall framing in middle

Roof framing on bottom

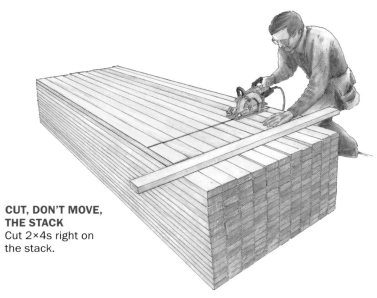

CUT, DON'T MOVE, THE STACK
Cut 2×4s right on the stack.

Acceptable Allowances

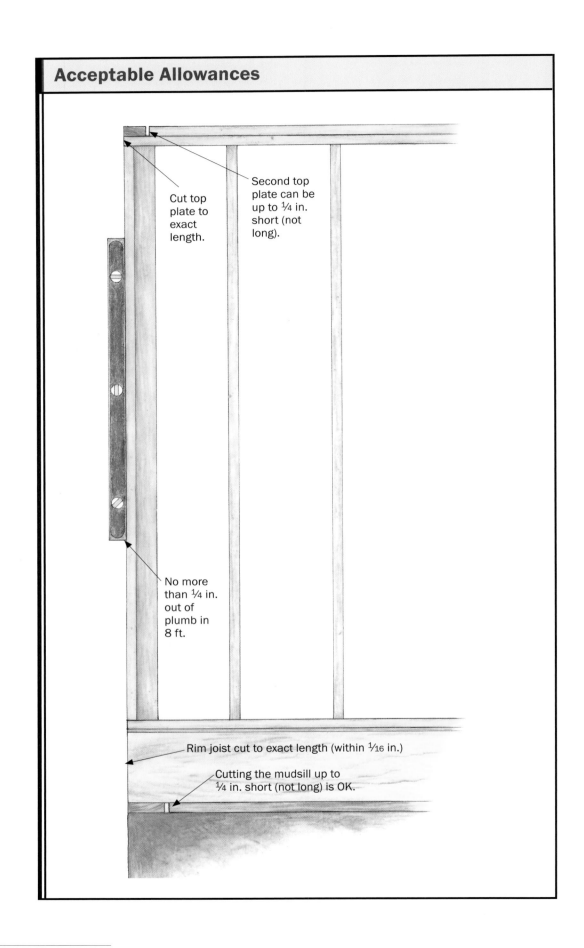

Cut top plate to exact length.

Second top plate can be up to 1/4 in. short (not long).

No more than 1/4 in. out of plumb in 8 ft.

Rim joist cut to exact length (within 1/16 in.)

Cutting the mudsill up to 1/4 in. short (not long) is OK.

2. Build a House, Not Furniture

In other words, know your tolerances. Rafters don't have to fit like the parts of a cabinet. Nothing in frame carpentry is perfect, so the question is this: What's acceptable?

You do need to get started right, and that means the mudsills. Whether they're going on a foundation or on a slab, they need to be level, straight, parallel, and square. But there's no harm done if they're cut ¼ in. short. A rim joist, on the other hand, needs to be cut to the right length (within ¹⁄₁₆ in.) before being nailed to the mudsill.

When it comes to wall framing, the bottom plate also can be ¼ in. or so short, but the top plate needs to be cut to exact length (again within ¹⁄₁₆ in.) because it establishes the building's dimension at the top of the walls. But the plate that sits on top of that, the cap or double plate, should be cut ¼ in. short so that intersecting walls tie together easily.

Once you've raised the walls, how plumb or straight is good enough? In my opinion, ¼ in. out of plumb in 8 ft. is acceptable, and a ¼-in. bow in a 50-ft. wall won't cause harm to the structure or problems for subcontractors. Take special care by framing as accurately as possible in the kitchens and in the bathrooms. These rooms require more attention, partly because of their tighter tolerances, but also because the work of so many trades comes together here.

3. Use Your Best Lumber Where It Counts

These days, if you cull every bowed or crooked stud you may need to own a lumber mill to get enough wood to frame a house. How do you make the most of the lumber that you get?

Use the straightest stock where it's absolutely necessary: where it's going to make problems for you later on if it's not straight. Walls, especially in baths and kitchens, need to be straight. It's not easy to install cabinets or tile on a wall that bows in and out. And straight stock is necessary at corners and rough openings for doors.

The two top wall plates need to be straight as well, but the bottom plate doesn't. You can bend it right to the chalkline and nail it home. If you save your straight stock for the top plates, you'll have an easy time aligning the walls. And every project needs lots of short stock for blocking; take your bowed material and cut it into cripples, headers, and blocks.

Take special care by framing as accurately as possible in the kitchens and in the bathrooms. These rooms require more attention, partly because of their tighter tolerances, but also because the work of so many trades comes together here.

PICK THE RIGHT STOCK FOR THE RIGHT PLACE
Straight lumber (left) is important for many locations, but some places, like the bottom wall plate, don't need perfectly straight stock.

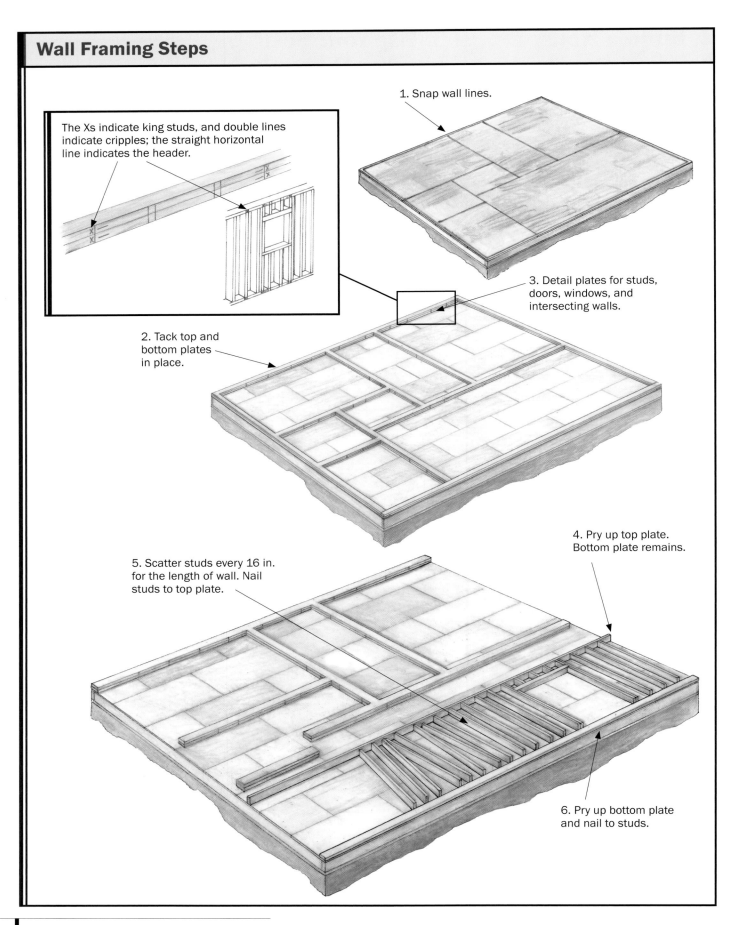

1. Snap wall lines.

The Xs indicate king studs, and double lines indicate cripples; the straight horizontal line indicates the header.

3. Detail plates for studs, doors, windows, and intersecting walls.

2. Tack top and bottom plates in place.

4. Pry up top plate. Bottom plate remains.

5. Scatter studs every 16 in. for the length of wall. Nail studs to top plate.

6. Pry up bottom plate and nail to studs.

4. Work in a Logical Order

Establish an efficient routine for each phase of work, do it the same way every time, and tackle each phase in its logical order. In the long run, having standard procedures will save time and minimize mistakes. Let's take wall framing as an example (see the illustration on the facing page).

First I snap all of the layout lines on the floor; then I cut the top and bottom plates and tack all of them in place on the lines. Next I lay out the plates, detailing the location of every window, door, stud, and intersecting wall.

I pry up the top plate and move it about 8 ft. away from the bottom plate, which I leave tacked to the deck. I scatter studs every 16 in. for the length of the wall. I nail the top plate to the studs and keep the bottom of the studs snug against the bottom plate. This helps to keep the wall square, straight, and in position to be raised. I try to establish a rhythm and work consistently from one end to the other. Once the top plate is completely nailed, I pry up the bottom plate and repeat the process on the bottom.

It's worth saying that I didn't just make up these steps; they evolved over time. Recognizing inefficiency is an important part of framing.

Plan for All the Pieces, Not Just Framing

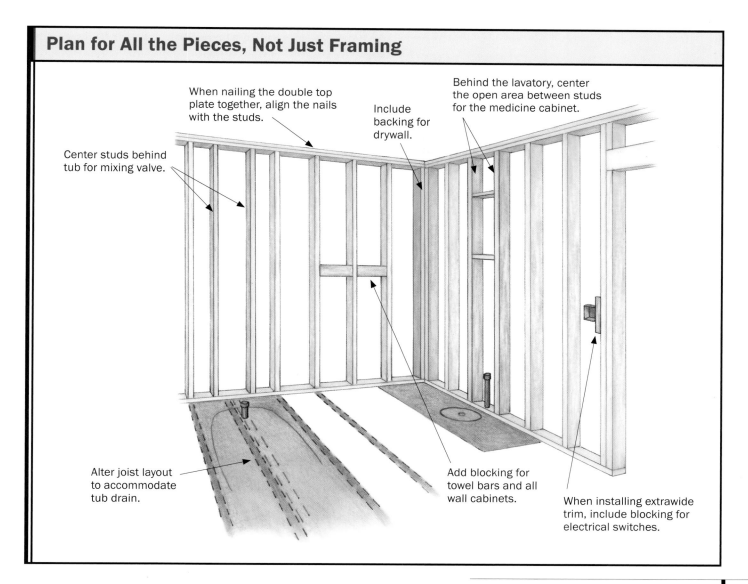

When nailing the double top plate together, align the nails with the studs.

Behind the lavatory, center the open area between studs for the medicine cabinet.

Include backing for drywall.

Center studs behind tub for mixing valve.

Alter joist layout to accommodate tub drain.

Add blocking for towel bars and all wall cabinets.

When installing extrawide trim, include blocking for electrical switches.

5. Keep the Other Trades in Mind

If you want to waste time and money when framing, don't think about the electrical work, the plumbing, the heat ducts, the drywall, or the finish carpentry. Whether you do them yourself or hire subcontractors, these trades come next. And unless you're working with them in mind every step of the way, your framing can be in the way (see the illustration on p. 11).

For example, when you nail on the double top plate, keep the nails located over the studs. This tip leaves the area between the studs free for the electrician or plumber to drill holes without hitting your nails.

6. Don't Measure Unless You Have To

The best way to save time when you're framing a house is by keeping your tape measure, your pencil, and your square in your nail pouch as much as possible. I have to use a tape measure to lay out the wall lines accurately on the deck, but after that, I cut all of the wall plates to length by cutting to the snapped wall lines. I position the plate on the line, eyeball it, and then make the cuts at the intersecting chalkline.

Another time-saver is to make square crosscuts on 2×4s or 2×6s without using a square. Experience has shown me that with a little practice, anyone can make these square cuts by aligning the leading edge of the saw's base, which is perpendicular to the blade, with the far side of the lumber before making the cut (see the illustration below).

Trimming ¼ in. from a board's length shouldn't require measuring. Ripping (lengthwise cuts) longer pieces also can be done by eye if you use the edge of the saw's base as a guide. Train your eye. It'll save time cutting, and as you develop, you'll also be able to straighten walls as easily by eye as with a string.

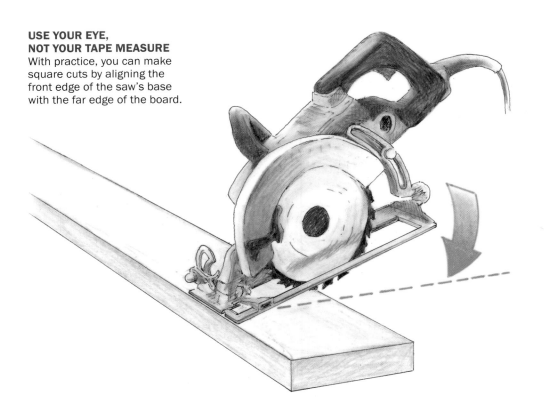

**USE YOUR EYE,
NOT YOUR TAPE MEASURE**
With practice, you can make square cuts by aligning the front edge of the saw's base with the far edge of the board.

7. Finish One Task before Going On to the Next

My first framing job was with a crew that would lay out, frame, and raise one wall at a time before moving on to the next. Sometimes they would even straighten and brace the one wall before proceeding. We wasted a lot of time constantly switching gears.

If you're installing joists, roll them all into place and nail them before sheathing the floor. Snap all layout lines on the floor before cutting any wall plates, then cut every wall plate in the house before framing. If you're cutting studs or headers and cripples, make a cutlist for the entire project and cut them all at once. Tie all the intersecting walls together before starting to straighten and brace the walls.

Finishing before moving on is just as important when it comes to nailing and blocking. You might be tempted to skip these small jobs and do them later, but don't. Close out each part of the job as well as you can before moving on to the next. Working in this way helps to maintain momentum, and it prevents tasks from being forgotten or overlooked.

8. Cut Multiples Whenever Possible

You don't need a mathematician to know that it takes less time to cut two boards at once than it does to cut each one individually.

If you have a stack of studs that all need to be cut to the same length, align one end of the top row, snap a chalkline all the way across, and cut the studs to length right on

You don't need a mathematician to know that it takes less time to cut two boards at once than it does to cut each one individually.

Save Time by Cutting More Than One

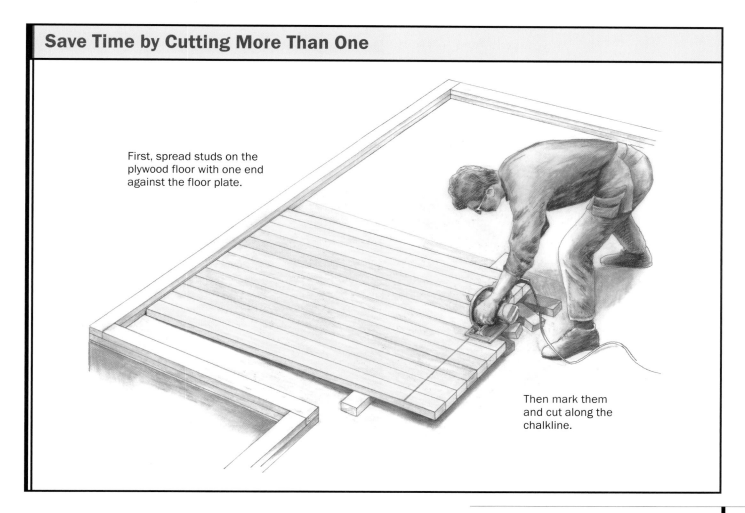

First, spread studs on the plywood floor with one end against the floor plate.

Then mark them and cut along the chalkline.

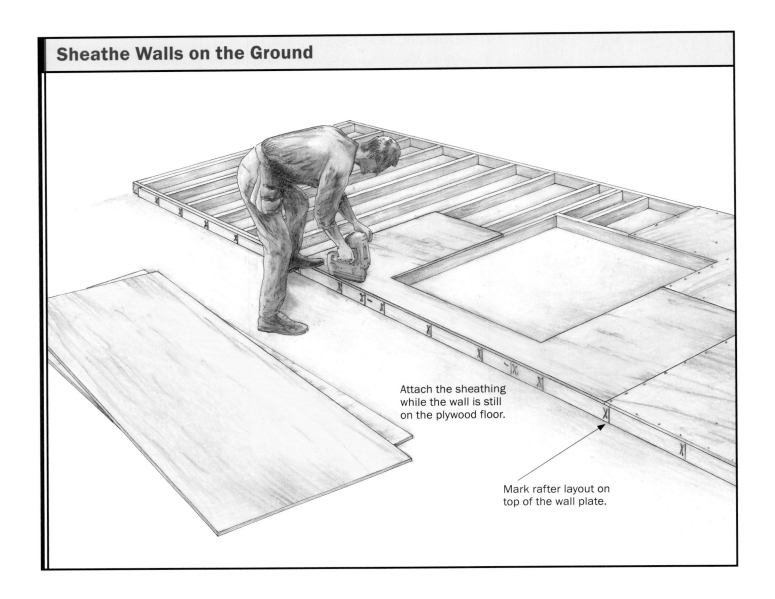

Attach the sheathing while the wall is still on the plywood floor.

Mark rafter layout on top of the wall plate.

the pile. Or you can spread them out on the floor, shoving one end against the floor plate, snap a chalkline, and cut them all at once.

Joists can be cut to length in a similar way by spreading them out across the foundation and shoving one end up against the rim joist on the far side. Mark them to length, snap a line, and cut the joists all at once (see the illustration on p. 13).

Also, don't forget to make repetitive cuts with a radial-arm or chop/miter saw outfitted with a stop block, which is more accurate and faster than measuring and marking one board at a time.

9. Don't Climb a Ladder Unless You Have To

I don't use a ladder much on a framing job except to get to the second floor before stairs are built. Walls can be sheathed and nailed while they're lying flat on the deck (see the illustration above). Waiting until the walls are raised to nail on plywood sheathing means you have to work from a ladder or a scaffold. Both are time-consuming.

With a little foresight, you can do the rafter layout on a double top plate while it's still on the floor. Otherwise, you'll have to move the ladder around the job or climb on the walls to mark the top plate.

10. Know the Building Code

Building codes exist to create safe structures. Because building inspectors are not capable of monitoring all parts of every project, it's your responsibility to know the building code and to build to it.

For instance, the code actually specifies how to nail a stud to a wall plate. You need two 16d nails if you're nailing through a plate into the end of the stud, or four 8d nails if you're toenailing. When you nail plywood or oriented strand board (OSB) roof sheathing, you need a nail every 6 in. along the edge of the sheathing and every 12 in. elsewhere. And if you're using a nail gun, be careful not to overdrive the nails in the sheathing.

A final word: If special situations arise, consult the building inspector. He or she is your ally, not your enemy. Get to know the building code for your area. Get your own copy of the IRC (International Residential Code) and build well, but build efficiently, with the understanding that perfection isn't what is required.

Larry Haun, author of The Very Efficient Carpenter *(The Taunton Press, Inc.) and* Habitat for Humanity How to Build a House *(The Taunton Press, Inc.), has been framing houses for more than 50 years. He lives in Coos Bay, Oregon.*

Work Safely Whatever the Rule

Working safely should be at the top of your priority list. Safety glasses, hearing protection, and a dust mask should be the norm, as should attention around coworkers or dangerous debris.

Safety devices and good intentions, however, won't help if your mind isn't on the work. Pay attention, approach the work with a clear head, listen to that inner voice that says, "This is too dangerous," and be extra careful toward the end of the day.

Roofing Code Requirements

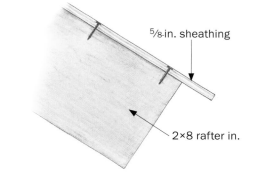

⅝-in. sheathing

2×8 rafter in.

Roof sheathing is nailed every 6 in. along the edges and every 12 in. elsewhere. In high-wind areas, sheathing along the eaves, rakes, and ridges is nailed every 6 in.

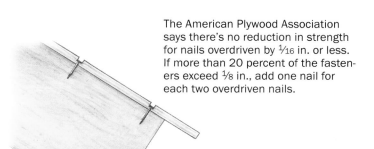

The American Plywood Association says there's no reduction in strength for nails overdriven by 1/16 in. or less. If more than 20 percent of the fasteners exceed 1/8 in., add one nail for each two overdriven nails.

Framing with a Crane

■ BY JIM ANDERSON

The other day, I kept an eye on the three-man framing crew working across the street. Two of the guys spent the day in the mud, hauling material for the second floor into the house. When my crew was rolling up for the day, the other crew had carried most of the lumber inside the house but still needed to pull it up to the second floor. It cost this crew about $300* in labor just to get the studs in the door.

I'd have hired a crane for this job and moved all that lumber in about an hour, which would have cost me $125. Even if there were no money savings, saving wear and tear on my body and those of my crew would make the crane worthwhile.

I started to think of better ways to use a crane six years ago when my brother and I went into business framing houses. We carried most of the material the hard way, but we hired a crane to set the steel I-beams.

It occurred to us that although the crane company charged us for a full hour of crane time, setting those three steel beams took 20 minutes. With that realization, we decided to fill the other 40 minutes of that hour by using the crane as our laborer (see the photo on the facing page).

My brother has moved on, and I now have my own two-helper crew. I still call in a crane several times for an average house. Where I work, in the suburbs south of Denver, Colorado, cranes are pretty common. The ones that I hire usually have no move-in fee, just a one-hour minimum charge of about $125.

Preparation Is Key

Before the crane arrives, I try to ensure that the lumber is dumped fairly close to the house, but not where it will be in the crane's way. When the crane does arrive, I discuss the sequence of the lift with the operator and crew. For efficiency, everybody has to know what's coming next.

One crew member stays near the material to rig it to the crane (see the sidebar on p. 18). The other two stay near where the material will be installed. To avoid confusion, one of these carpenters is the designated signaler (see the sidebar on pp. 20–21), whereas the other jockeys the load into position.

Most of the houses that my crew and I frame have three to five steel beams holding

A Well-Rigged Job Goes Smoothly and Safely

Three kinds of rigging equipment get us through any house. The most frequently used are nylon straps, followed by steel cables and four-chain rigging.

Nylon Straps Are Rigged in Two Ways

Cradle-rigging runs the strap under the load in a U-shape. Great for trusses and walls, they should never be used to lift studs or joists overhead.

Choke-rigging tightens as the crane lifts. Choke rigs limit movement within lifts such as studs or sheathing, and they keep beams from slipping out of single straps. To avoid excessive flexing, I-joists (at right) are choked in about one-quarter of the joists' length from each end.

Cables and Four-Chain Rigs Are Less Common

Four-chain rigs raise the roof. Hooked to the crane with a ring, each chain can be snubbed to a length that will enable a preassembled truss rack to fly level.

Steel cables can damage the edges of material. They're limited to raising single trusses and sliding under material dropped directly onto the ground so that it can be raised enough to get a strap below.

up the first floor. These beams are the first things that the crane sets (see the photos below). We either have the Lally columns cut to length or have ready temporary posts. Long 2×4s are on hand to brace the I-beam to the mudsills once it's in place.

We set the beam that's farthest back in the building first, then move sequentially to the front. This process avoids swinging anything over set beams, eliminating the chance of dropping one beam and taking out two.

We stack any scrap from the basement on a piece of sheathing. We lift this scrap out of the hole with the crane and swing it right over to the Dumpster®.

Any built-up wood or laminated veneer lumber (LVL) beams are nailed together, and any necessary hangers are installed beforehand. As we work back to front, I set these beams in sequence with the steel.

The stacks of floor joists are next (see the top right photo on the facing page), and after they have all been set on the foundation walls, we move as much of the material as we can closer to the house. We put the lifts of floor sheathing on top of 2×4 stickers about 3 ft. from the front of the house, allowing room to work but putting it within easy reach. Then we use the crane to move the rim-joist material to the top of the stacks of sheathing. We cut the rims here, using the sheathing stacks as 1,500-lb. sawhorses.

While You're Setting the Steel, Have the Garage Walls Ready to Lift

Most of the houses we frame have three-car garages with one single door and one double door. A typical garage-door wall is 30 ft. long and 10 ft. tall, with an 18-ft. and a 9-ft. double-LVL header (see the top photos on p. 20). I never want to lift a wall like this by hand. Before the crane arrives, we frame and stand

The steel is ready to go when the crane arrives. The wood sills to which the joists will be nailed are already attached to the steel, and the joist locations are marked.

the sidewalls that we can easily lift by hand. We also frame the front wall but leave it on the ground for the crane to lift.

The crane lifts the wall and swings it to the garage foundation. As it gets close, we guide the anchor bolts into the holes we've drilled for them in the bottom plate, then down the wall. Now the crane holds the wall in place as we nail the corners, tie in the plates, and nail on some braces. With the wall set and braced, we give the "all finished" signal and send the crane home. This entire lift—I-beams to garage wall—usually takes only about an hour.

A crane makes short work of raising top-heavy garage walls. Once the wall is set over the anchor bolts, the crane steadies the wall until it's tied to the others and braced.

Getting the Crane to Do What You Need

There are standard hand signals that all crane operators and the people who hire them should know. In addition, three rules and a suggestion can make communication a sure thing.

1. Keep your signal in one place.

2. If you can't see the operator through a maze of studs, trusses, or bracing, the operator can't see you. Make eye contact, and then make your signals in front of your face.

3. If your gloves and clothing are of similar colors, make your signals away from your body where the operator can see. Another way to communicate with the operator is with two-way radios.

To raise or lower a load, point up or down and rotate your finger. To move the load slowly, put your opposite hand above or below your signal, as if you're pointing at your palm. When the load is down, a quick circle with the hand signals that all is clear and that the hook can be dropped to free the load.

Lifting with a Crane Beats Carrying Material to the Second Floor

After we've framed the first-floor walls, we build a section of the second floor to serve as a staging area for our next lift. This area is usually a corner that's, say, 500 sq. ft. to 800 sq. ft. (see the photo at right). We also frame most of the walls around this area before placing any material here. This prep work saves having to move lumber to make room to frame the walls.

Once this staging area is done, I call the crane again to lift the studs, the plates, and the sheathing for the second-floor walls, as well as the balance of the second-floor joists and beams.

Full lifts of studs or of OSB sheathing are pretty heavy loads that should never be set

Be prepared for the crane's second visit. This second-floor staging area is framed to provide storage space for crane-lifted studs and sheathing. Because stored material would otherwise be in the way, the author also frames the walls before the crane arrives.

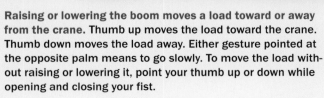
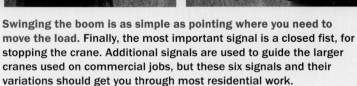

Raising or lowering the boom moves a load toward or away from the crane. Thumb up moves the load toward the crane. Thumb down moves the load away. Either gesture pointed at the opposite palm means to go slowly. To move the load without raising or lowering it, point your thumb up or down while opening and closing your fist.

Swinging the boom is as simple as pointing where you need to move the load. Finally, the most important signal is a closed fist, for stopping the crane. Additional signals are used to guide the larger cranes used on commercial jobs, but these six signals and their variations should get you through most residential work.

in the middle of a joist span. We set lifts of material on 2×4s to spread the weight and to leave room to remove the straps once the crane lets go of the load. We always set our material on or near the main bearing beams and walls below the floor. If for some reason a load must be set midspan, we split it into smaller bundles that can be spread out. We sometimes build a temporary wall below the floor to help spread the load to more joists.

During this lift, we set any second-floor steel or built-up LVLs. Commonly, there is a beam that runs between two walls, each end supported by posts made of studs or by Lally columns enclosed within pockets in the wall. Although we frame these walls with their top plates uninterrupted, we leave the posts out until we set the beam.

To set a beam that runs between walls, we pull down one end and slide it far enough into its pocket that the beam's other end clears the wall plate. Then we have the crane operator lower the beam, and we seesaw it into place (see the bottom photo on p. 19). With the crane snugging the beam to the underside of the top plates, we tip in the supporting columns.

Raising Balloon-Frame Walls

The only first-floor walls we don't lift by hand are any tall, balloon-frame walls, such as those of rooms with very high ceilings (see the photo below left). We frame and sheathe these walls with the other walls but

With the walls partially raised and shored for safety, the braces are tacked so that they can pivot as the wall rises.

Two straps are used to raise tall walls. Had there been no windows, the crew would have cut small holes in the sheathing for the straps.

leave them flat until the crane comes again. For safety alone, lifting these walls is worth hiring a crane.

With the wall's bottom plate on its layout line, we secure it to the deck about every 4 ft. using pieces of the steel strap that bands lifts of lumber. This strap acts as a hinge, keeping the wall from kicking out during the lift.

We stop the lift about a third of the way up to attach the braces that will steady the raised wall (see the right photo on the facing page). These braces go about two-thirds of the way up the wall. Once the braces are nailed to the wall, we stand it up the rest of the way, then plumb and brace it before unhooking the crane.

The Final Lift Sets the Roof Trusses

With the second-floor walls plumbed and lined, it's time for the crane again. This time it will lift our sheathing, roof-framing lumber, prebuilt truss racks (see the top right photo on p. 18), and any single trusses. First, we swing up the roof sheathing and set it in three or four spots on the second floor. Then we send up the lumber for framing dormers and valleys. This lumber is usually set in the main hallway of the upper level, where we have room to maneuver long pieces up into the roof framing.

When the site and truss design allow, we assemble, sheathe, and brace the trusses into 6-ft. to 18-ft. roof sections on the ground (see the bottom right photo on p. 18). We can set a simple gable roof, preassembled into two to three sections, in about a half-hour.

Our total crane time on an average house is about 4½ hours. In most cases, we hire a 3-ton crane. These cranes have 90-ft. booms that will reach about 60 ft. with most loads that we see in residential construction. Only once have we needed a larger crane. If you aren't sure of the size crane that you need, call the crane company and describe your lift. They'll know.

Crane Safety Is Mostly Common Sense

Cranes offer immediate safety benefits by lifting heavy loads that you might otherwise attempt to manhandle. And by reducing the repetitive toting and lifting of construction, they can foster long-term benefits in the form of healthy backs, knees, and shoulders.

However, cranes bring with them some danger, simply because they carry heavy things overhead. Obeying three rules should get you home safe after every day of crane work.

- Wear a hard hat: OSHA requires it.
- Don't stand under suspended loads.
- Don't become trapped between the load and a wall or a drop-off. A sudden horizontal movement of the load could crush you or send you flying.

There are two other advantages to hiring a crane. The first is safety. We once helped another framer attempt to raise a large wall by hand. It got away from us, pinning one of my employees below. Added to his pain and disability is the fact that my worker's comp premiums went up 50 percent.

The second advantage is job-site security. Using a crane to lift material into the house as soon as possible makes it less accessible to thieves. They'll likely head across the street to the house where the lumber is still sitting where it was delivered, conveniently right next to the curb.

Prices are from 2001.

Jim Anderson is a framing contractor living in Littleton, Colorado.

All about Headers

■ BY CLAYTON DEKORNE

Like many carpenters in the Northeast, I was taught to frame window and door headers by creating a plywood-and-lumber sandwich, held together with generous globs of construction adhesive and the tight rows of nails that only a nail gun could deliver. Years later, I learned my energetic efforts to build a better header were an exceptional waste of time and resources. Neither the plywood nor the adhesive contributed much strength, only thickness, and this perfect thickness helped only to conduct heat out of the walls during the severe winters common to the region.

At the same time I was laying up lumber sandwiches, young production framers on the West Coast were framing headers efficiently using single-piece 4×12s. They needed only to be chopped to length and filled the wall space above openings, eliminating the need for maddeningly short cripple studs between the top of the header and the wall plate. Nowadays, however, such massive materials are relatively scarce and remarkably expensive, even on the West Coast. So although solid-stock headers certainly save labor, they no longer provide an economical alternative.

With these experiences in mind, I set out to discover some practical alternatives, sur-veying a number of expert framers in different regions of the country. Header framing varies widely from builder to builder and from region to region. Even when factors such as wall thickness and load conditions are made equal, building traditions and individual preferences make for a wide range of header configurations. The examples shown here are just a few of the options possible when you mix and match features, notch cripple studs, and sift in engineered materials. But they aptly demonstrate a number of practical considerations that must be kept in mind when framing a good header.

Big Headers Need More Studs

A header transfers loads from the roof and floors above to the foundation below by way of jack studs (see the sidebar on p. 26). This means the header not only must be deep enough (depth refers to the height of a beam: 2×10s are deeper than 2×6s) for a given span to resist bending under load, but also must be supported by jack studs on each end that are part of a load path that continues to the foundation.

TIP

Although most carpenters tend to think that more nails are a sign of good workmanship, headers need to be nailed with only one 10d common nail every 16 in. along each edge.

The header you choose. This decision affects not just strength, but also cost, energy efficiency, and drywall cracking.

Like a Bridge over a Ravine, a Header Spans a Window or Door

Headers are short beams that typically carry roof and floor loads to the sides of openings for doors or windows. Jack studs take over from there, carrying the load to the framing below and eventually to the foundation. That's called the load path, and it must be continuous. The International Residential Code (the most common code nationwide) has a lot to say about headers, including the tables you need to determine the size header required for most situations. If it's not in the IRC, you need an engineer.

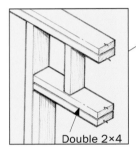

Double 2×4

WHERE YOU DON'T NEED A STRUCTURAL HEADER

In a gable-end wall that doesn't support a load-bearing ridge, or in an interior nonbearing partition, headers up to 8 ft. in length can be built with the same material as the wall studs. Inside, a single 2× is often sufficient, although a double 2× is helpful for securing wide trim. In a gable-end wall, however, a double 2× is needed to help resist the bending loads exerted by the wind.

Load path must be continuous from roof to foundation.

CRIPPLE STUDS FILL THE VOID

If the header doesn't fill the space all the way to the top plate, cripple studs are used to carry the load from the rafters, joists, or trusses above to the header below. For nonbearing headers, the IRC requires no cripples if the distance to the top plate is less than 24 in.

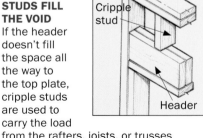

Cripple stud

Header

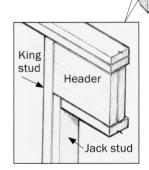

King stud

Header

Jack stud

KINGS HIGH, JACKS LOW

King studs are the same height as the wall studs, running plate to plate. Nails driven through them into the header's end grain stabilize the header. Jack studs are shorter and fit below the header to carry loads downward. Because longer-spanning headers usually carry greater loads, you may need an extra jack under both ends of big headers. Check your building code.

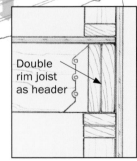

Double rim joist as header

HIDDEN HEADER

Here's a wood-saving trick. By simply adding an additional member to it, the rim joist above an opening can serve as a header. Two caveats: You'll need to use joist hangers to transfer the load to the header. And if a standard header of this size requires double jack studs, so does a double rim joist.

number of jack studs for most common situations. Although most windows and doors require just one jack stud at each end, long spans or extreme loads may call for two or more jack studs to increase the area bearing the load. If the loads on any header are concentrated over too small an area, the wood fibers at the ends of the header can be crushed. This can cause the header to drop, which in turn can crack drywall or, particularly with patio doors and casement windows, cause the door or window to jam.

Header hangers, such as the Simpson Strong Tie® HH Series (see the sidebar on p. 32), can be used to eliminate jack studs altogether. I've used them in some remodeling situations when I needed to squeeze a patio door or a wide window into an existing wall that didn't have quite enough space for double jack studs. One jack and a Simpson HH Series hanger did the trick.

Continued on p. 32.

Installation Guidelines

Typically, header height is established by the door height, and window headers are set at this same height. In homes having 8-ft. ceilings, a header composed of 2×12s or of 2×10s with a flat 2×4 or 2×6 nailer on the bottom accommodates standard 6-ft. 8-in. doors, as shown in the illustration below.

In a custom home with cathedral ceilings and tall walls, however, header heights can vary widely. And if the doors are a nonstandard height, you'll need to figure out the header height. Finding the height of the bottom of the headers above the subfloor is a matter of adding up the door height, the thickness of the finished-floor materials, and 2½ in. (to allow space for the head jamb and airspace below the door). There are exceptions. Pocket doors typically require a rough opening at least 2 in. higher than a standard door. Windows may include arches or transoms, which affect the rough opening's height.

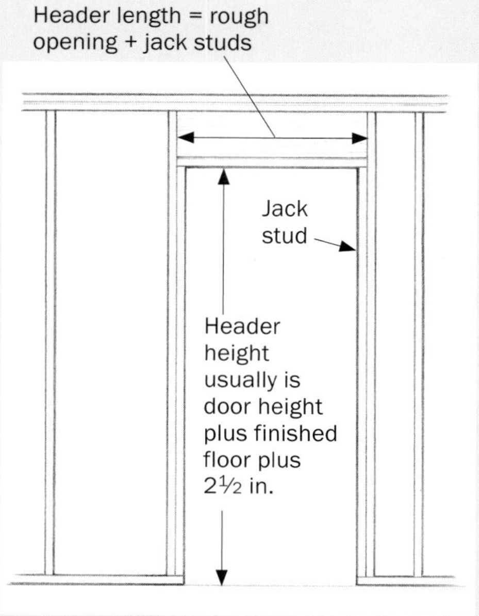

To find the header length for windows, add 3 in. to the manufacturer's rough-opening dimension if there is to be one jack stud on each side, or 6 in. if two jacks are called for. For doors with single jack studs, add 5½ in. to the door width to allow for jack studs, door jambs, and shim space. If double jacks are needed, then the header should be 8½ in. longer than the door width.

These guidelines follow one fundamental rule of framing rough openings: Know your windows and doors. If you don't have the window or door on site, at the very least check the manufacturer's catalog to verify the rough-opening dimensions. Don't rely on the plans alone, and when in doubt, call the manufacturer.

Header length = rough opening + jack studs

Jack stud

Header height usually is door height plus finished floor plus 2½ in.

Sawn-Lumber Headers

When it comes to sawn-lumber headers, traditional materials still carry the load.

Double 2×6 Header

Fine Homebuilding contributing editor Mike Guertin, whose day job is building houses in Rhode Island, uses the smallest allowable header depth to span the opening. Although he must toenail cripples above each header, he argues that this header is the most economical. For starters, it conserves lumber. It also reduces the area of solid material in the wall, thus reducing thermal bridging. Although the area is kept to a minimum, Guertin is also careful to keep the header to the outside of the wall, providing a gap that may be insulated with foam or wet-spray cellulose when the rest of the wall is insulated. A 2×3 nailed to the lower edge of the header provides attachment for trim.

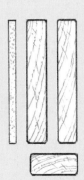

Double 2×10 Header

A common header variant is used by North Carolina builder John Carroll. Built from double 2×10s, a stud-width nailer flat-framed along the bottom edge eases attaching sheathing or trim. Because this header is less than the full thickness of the wall, it allows for a piece of ½-in. foam to add a bit of insulation.

Insulated Header

Custom builder David Crosby of Santa Fe prefabs insulated headers from 2×10s and 2-in. extruded polystyrene foam. This option works particularly well in the cold mountains of northern New Mexico, where air temperatures can fall well below zero on winter nights. Even adding some ½-in. foam to a double header in a 2×4 wall improves the thermal performance. Although lumber in New Mexico is typically quite dry due to the arid climate, Crosby ties the header to the jack stud with metal framing plates to control header shrinkage that could open gaps in the trim.

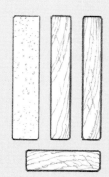

Built-Up Plywood and Lumber

In this header sandwich, plywood adds only thickness so that the header will fit flush to each face of the wall. There is little strength added, even if the header is spiked together with construction adhesive between each layer. Construction adhesive adds nothing to the strength of a beam.

Before assembling this (or any other header), crown the lumber, marking it clearly with a lumber crayon, and keep the crown up. Rip plywood ½ in. narrower than the lumber to prevent the pieces from hanging over the edges, especially if the lumber has a crown.

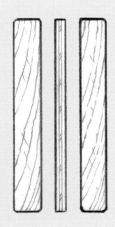

Solid-Stock Header

Once standard fare for West Coast production framers, a solid header made with a single 4×12 tucks tight under the top plates in a wall, eliminating the need for short cripple studs. Although this option saves substantial labor, the availability of full-dimension lumber is limited mainly to the West Coast. Even there, solid-stock headers are expensive and may not be cost-effective unless the opening requires the load-bearing capacity of such large-dimension stock.

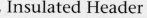

Engineered-Wood Headers

Engineered-wood headers cost more, but they do more, too.

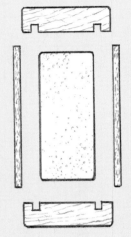

Store-Bought Insulated Headers

Essentially a double-webbed I-joist with a chunk of rigid foam wedged in the middle, these engineered SWII Headers from Superior Wood Systems® offer insulation, strength, and light weight. You may have a hard time finding them locally, though, because they're new enough that distribution varies regionally. Price varies as well, depending on freight costs and markup. Hammond Lumber in Bangor, Maine, sells 14-ft. long, 5½-in. by 11¼-in. SWII Headers for about $90.

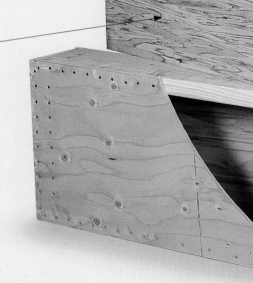

Parallel-Strand Lumber

Parallel-strand lumber, such as TrusJoist® MacMillan's Parallam®, is available as stud-width stock. Performing much like LVL, parallel-strand header stock is pricier than solid sawn lumber but 1½ times as stiff and 3 times as strong.

Laminated-Veneer Lumber

Engineered lumber, shown in this header made from two pieces of 1¾-in. by 16-in. LVL, offers some advantages over sawn lumber. Although it's more expensive for smaller headers, engineered lumber is available in depths that can span distances sawn lumber simply isn't up to. And it's typically more stable, resulting in fewer drywall cracks.

Structural Box Beam

A box-beam header is a viable way to site-build long-span headers. A technical bulletin, *Nailed Structural-Use Panel and Lumber Beams,* outlines the design and fabrication of these stud and plywood beams. Because they end up being thicker than the studs, these plywood beams are better suited for long-span headers in an unfinished garage, where the exact thickness is a slight concern. For a 2×6 wall, though, you can make a box beam using 2×4 blocking and nominal ¾-in. structural plywood. A ½-in. furring strip brings such headers to the full wall thickness. And they can be stuffed with insulation.

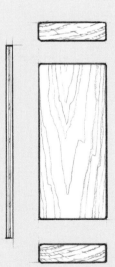

Sometimes, particularly in remodeling, there just isn't room for a jack stud. The IRC permits header hangers, such as Simpson's HH4 for 2×4 walls and HH6 for 2×6 walls, to substitute for single jack studs. These hangers are spiked with 16d common nails to the king stud.

Hangers eliminate jack studs.

How Big a Header Do You Need?

Unless you're an engineer, the easiest way to size headers built with dimensional lumber is to check span charts, such as those in the IRC. The old rule of thumb is that headers made of double 2× stock can span safely in feet half their depth in inches. So by this rule, a double 2×12 can span 6 ft.

However, header spans vary not only with size, but also with lumber grade and species, with the width of the house, with your area's snow load, and with the number of floors to be supported. Consequently, the IRC provides 24 scenarios in which that double 2×12 header can span a range from 5 ft. 2 in. to 9 ft. 9 in. Check the code.

The Trouble with Cripples

Header size often is based on factors other than strength requirements. Many framers purposely oversize headers to avoid filling the space between the header and the double top plate with short studs (cripple studs, or cripples). In a nominal 8-ft.-tall wall, a typical cripple stud measures 6 in. to 7 in. Such short studs are ungainly and are prone to splitting when they are nailed in place. Yet a double 2×12 header can be tucked beneath the double top plate, filling this miserable space and creating a proper opening for common 6-ft. 8-in. doors. Alternatively, builder John Carroll relies on a double 2×10 header with a 2×6 nailed flat along the bottom edge, which provides nailing for the head trim in a 2×6 wall (see the sidebar on pp. 28–29).

However, such deep headers are oversize and add considerable cost, not to mention waste wood. Most window and door openings are only 3 ft. or so and might only require 2×6 headers. But perhaps the biggest drawback of wide lumber is that there's more of it to shrink. Framing lumber may have a moisture content of 19 percent. Once the heat is turned on, lumber typically dries to a moisture content of 9 percent to 11 percent, shrinking nominal 2×10s and 2×12s as much as ¼ in. across the grain. On the other hand, 2×6s might shrink only half that.

Shrinkage reduces the depth (or height) of the header; because the header is nailed firmly to the double top plate, a gap usually opens above the jack studs. As the header shrinks, it tends to pull up the head trim, which has been nailed to it, opening unsightly gaps in the casing and cracking any drywall seam spanning the header. The gap above the jack stud now means the header isn't supporting any load—until the first wet snowfall or heavy winds bring a crushing load to bear on the wall and push the gap closed, causing the top plates to sag, which can crack the drywall in the story above.

Shrinkage can be reduced using drier lumber, preferably at about 12 percent. However, lumber this dry may be difficult to find unless you can condition it yourself. As an alternative for spanning a large opening, consider using engineered materials (see the sidebar on pp. 30–31). Laminated-veneer lumber or parallel-strand lumber (PSL) shrinks much less than ordinary lumber.

If wide dimensional lumber is unavoidable, structural engineer Steve Smulski suggests that cracking can be minimized by not fastening the drywall to the header. This way, the header moves independently of the drywall, which then is less likely to crack. To prevent trim from moving as the header shrinks, attach the top piece of trim to the drywall only, using a minimal number of short, light-gauge finish nails and a bead of adhesive caulk.

Avoiding Condensation

In cold climates, uninsulated headers can create a thermal bridge. According to Smulski, the uninsulated header makes the wall section above windows and doors significantly colder than the rest of the wall. When the difference between the inside and outside air temperatures is extreme, condensation may collect on these cold surfaces, and in the worst cases, mold and mildew may begin to grow.

To avoid condensation, it's important that any uninsulated header doesn't contact both the sheathing and the drywall. Unless you're building 2×4 exterior walls using full-thickness headers such as solid lumber or ones built out to 3½ in. with plywood, avoiding this situation is simple. Keep the header flush to the outside of the framing so that it contacts the sheathing. Because most other types of headers are narrower than the studs, there will be some airspace between the header and the drywall, which makes a dandy thermal break. In cold climates, a 2×10 insulated header, like the one used by David Crosby of Santa Fe, New Mexico, works well (see the sidebar on pp. 28–29). Another option that avoids solid lumber is a manufactured insulated I-beam header (see the photo on pp. 30–31).

Prices are from 2004.

Clayton DeKorne *is a carpenter and writer in Burlington, Vermont. He is the author of* Trim Carpentry *and* Built-Ins *(The Taunton Press, Inc.).*

Sources

APA Engineered Wood Association
Nailed Structural-Use Panel and Lumber Beams
Available online www.apawood.org

Simpson
4637 Chabot Dr., #200
Pleasanton, CA 94588
800-999-5099
www.strongtie.com

Superior Wood
1301 Garfield Ave.
Superior, WI 54880
800-375-9992
www.swi-joist.com

TrusJoist MacMillan's Parallam
www.ilevel.com

Fast and Accurate Framing Cuts

■ BY LARRY HAUN

Have you ever watched a professional chef chop vegetables? A blur of motion, the staccato sound of a knife on a cutting board, and a mound of neatly sliced carrots appears almost instantly. I always look for a fingertip among the carrots.

Some people have that same reaction when they watch me cut framing lumber—impressed by the speed, worried about safety. Like knives, circular saws are dangerous tools, but in experienced hands, a saw can be safely pushed to its limit.

I bought my first circular saw in 1951—a used worm drive for $85. I thought I'd died and gone to heaven. Suddenly, houses that used to take a week to frame could now be framed in a day.

But as the postwar demand for houses increased, so did the competition to build them. Surviving as a framing contractor

meant more than just doing good work; it meant working fast, too. In the quest for efficiency, I began to depend less on my measuring tape and more on marking and cutting boards in place, less on my square and chalkline, more on my eye.

Just as ear training helps a musician to play an instrument, I trained my eye to help guide sawcuts. The saw became an extension of my arm, sort of like a bat in the hands of a professional baseball player.

The Saw Base Lines Up the Cut

It takes literally a second to cut through a 2×4. But if you have to pull out a square and a pencil to mark a line to guide the cut after measuring and marking the length of the cut, and then put these tools back

The methods described here are not for beginners— they're for veteran carpenters, already comfortable with saws and ready to challenge themselves to work faster and more efficiently.

in your nail bag, the same cut takes you many times longer. Cutting square without drawing a line depends on the basic and oft-overlooked premise that the saw base is a rectangle and that the side of the base is parallel to the blade.

For a square cut on a 2×4, just line up the front edge of the saw base parallel with the far edge of the board. The blade should now be perpendicular or square to the board, and a cut in that position should also be square (see the photo at left).

For wider boards, hold the guard up out of the way, tip the saw forward, and line up

Cut wherever you're standing. With the length marked and the saw base aligned with the edge of the board, a 2×4 can be cut square in a second. Using your foot as a support saves having to move the lumber to a pair of sawhorses.

A three-second crosscut. Make a plunge cut with your blade on the mark and the saw base square to the board (photo above). With the saw running, lift the board so that the saw's weight helps to finish the cut (photo right).

the saw base with the edge of the board (see the bottom left photo on the facing page). Now plunge through and lift the near edge of the board so that the weight of the saw helps you to finish the cut (see the bottom right photo on the facing page). The whole process still takes around a second and with practice becomes one fluid motion. Although 2×4s or 2×6s can be cut safely with the board resting on your foot, the safest way to cut a wider board is to rest it on a 2× block.

The same process is used when cutting a 4×4 post or header that is too thick for the saw to cut in one pass, as seen in the photo sequence on this page. Once the first cut is made square, the saw kerf guides the next two cuts.

Perfect post cuts in a snap. Cut side one as you would a 2×4 (photo right). Next, rotate the 4× on edge and cut side two (photo below left). Keeping the blade in the kerf, roll the 4× forward and plunge through side three in chopsaw fashion (photo below right). The saw kerf from the first square cut aligns the next two.

Flip the board for safe short rips. To rip short boards, plunge in about halfway down the board (photo above) and rip to the end. Flip the board and reinsert the blade into the kerf to finish the rip. Raise the board at the end of the cut to let the saw clear the material (photo left).

Saw base can help gauge ripping widths. Knowing the distance between the blade and the edge of the saw base lets you rip many widths without a line. Here, the author has estimated ½ in. beside the base for a 2-in. rip.

Let the Saw Base Be the Rip Guide

I've also learned to trust my eye when ripping boards. The blade on my saw is 1½ in. from the left edge of the saw base and 3½ in. from the right edge. So it is easy to rip 2× or 4× widths just by running the saw base along the edge of the board. The chalkline stays in my pouch.

Almost anyone can come pretty close to guessing at ½ in. So the next step is ripping 2 in. or 4 in. widths by leaving ½ in. of stock showing along the edge of the board (see the photo at left). Next, try stepping in 1 in. for a 2½-in. cut and so on. The trick is to trust your eye.

I rip long lengths by elevating one end of the board and sawing downhill. When ripping short stock, like a 3-ft. section of 2×, I make the cut in two steps so that I can hold one end without putting my hand in danger

(see the photos above). This process involves plunge-cutting, so watch out for kickback (see the sidebar on the facing page).

Measure Twice, Cut Once, or Don't Measure at All

If you're eliminating the step of drawing or snapping lines to save yourself some time, imagine how quickly framing would go if you didn't have to measure. It may be unthinkable to some, but there are many times when I skip the step of measuring.

One such time is cutting floor joists in place (see the top photo on the facing page). Here, the cut joist falls neatly down onto the sill plate ready to be installed while the saw rides safely on the waste that rests on the rim joist. Simply keep the saw square to the stock and aim the blade to pass just inside the edge of the rim. Once the floor sheathing is nailed down, I snap lines on the floor locating all the walls. Now I can cut the wall plates just by laying them on top of the lines and cutting them in place. The walls are all cut, assembled, and raised—and I don't have to take my measuring tape out of its pouch.

Another occasion when you can cut without measuring first is when you are letting in a wooden brace. In these cases, the 1× brace acts as a template for the cut (see the bottom photos at right). First, lay the brace across the stud wall at 45 degrees. Make the plunge cuts beside the brace into each stud and plate. Then remove the wood between the two cuts by turning the saw over on its side and plunging through. The weight of the saw as it is on its side will make the blade want to bind in its kerf. So hold the saw carefully as you plunge in, and make sure you brace your elbows against your knees to resist any kickback.

Larry Haun, author of The Very Efficient Carpenter *(The Taunton Press, Inc.) and* Habitat for Humanity How to Build a House *(The Taunton Press, Inc.), has been framing houses for more than 50 years. He lives in Coos Bay, Oregon.*

Skip the tape, and cut the lumber in place. Here, the rim joist determines the length of the joist stock. After the saw is aligned with the rim joist, each joist is cut and falls on the sill, ready for installation.

Lumber acts as a template for a let-in brace. With the brace laid across the studs, the author braces the saw for possible kickback and plunges down beside the brace at each stud (photo left). A plunge from the side of the stud removes the wood to let in the brace (photo below).

Kickback

Kickback is bound to happen to you if you use a circular saw. I have a nine-stitch scar in my left leg as a reminder of that fact. Kickback occurs when the sawblade gets pinched in the kerf and the power of the motor forces the saw backward. This can be scary, so here are some guidelines for avoiding kickback.

- Make sure the blade guard works smoothly.
- Use both hands to guide the saw when necessary.
- Keep your body, especially your hands, out of the path of any potential kickback.
- Use a sharp blade with enough set to cut a kerf wider than the blade.

- Set the blade 1/8 in. deeper than the material to be cut.
- Cut in a straight line. Don't force or twist the saw as you cut.
- Let the cutoff end of 2× stock drop free.
- Support sheet goods and long stock on both sides of the cut to prevent sagging, which can pinch the blade.
- When kickback occurs, release the trigger and allow the saw to stop.

I find heavier saws, such as worm drives, less prone to kickback. I refuse to use some of the cheap, lightweight home-owner models. Saws with the handle in the back seem easier to control than those with the handle near the top.

Nailing Basics

■ BY LARRY HAUN

When I was a small boy growing up on the Nebraska high prairies during the Depression, my first hammer was a 16-oz., wood-handle model with one of its curved claws missing. It wasn't much good for pulling nails, but at the time it was all we could afford.

Our nails were gleaned from the site where an old barn had burned to the ground. I'd find round nails as well as the cut, square type and used to straighten bunches of them getting ready for childhood building projects. We'd bang together forts and playhouses made of wooden orange crates and stray boards. They were fun to build, but they didn't stand up very well to the strong, gusty winds that blew down from Wyoming.

In 1948 I had my first job as a paid carpenter. For a dollar a day, I helped an old craftsman put together a precut house. The house was shipped by rail and hauled to the site on a horse-drawn wagon. Over the next year or so, we put it together piece by piece. I was still using a curved-claw hammer, but this one had all its parts.

When the postwar building boom hit, I found myself in California with a 20-oz. straight-claw hammer in my hand. The demand for housing was so great that instead of building one house a year, we built 500. In the four decades since then, pneumatic and propane-powered nail guns have changed the way most nails are driven. But even with these quick and efficient tools, carpenters carry hammers in their nail pouches, and learning to use them is step one.

Hammers Are a Matter of Personal Preference

Early framers used the Plumb rigging hatchet for a hammer. Its perfect balance and long, comfortable handle made the hatchet easy to use for long days of constant nailing. But without claws, the hatchet was no good at pulling nails. So carpenters cut off the blade of the hatchet and welded on a set of claws. Guys started making these new well-balanced wood-handle framing hammers and selling them from the trunks of their cars on job sites. Word spread, and demand grew to the point where beauties such as the Dalluge® and Hart framers became favorites of carpenters all over the country.

The Art of Nailing

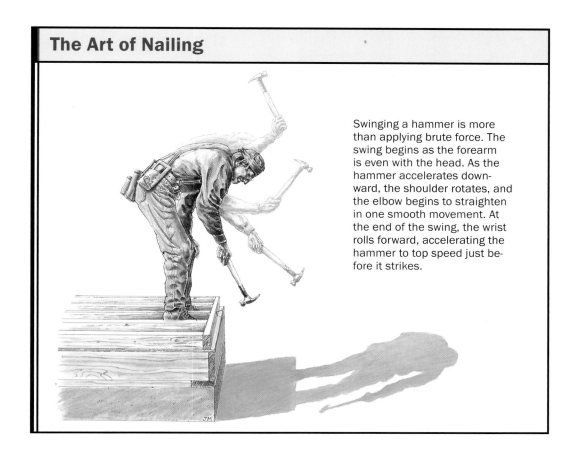

Swinging a hammer is more than applying brute force. The swing begins as the forearm is even with the head. As the hammer accelerates downward, the shoulder rotates, and the elbow begins to straighten in one smooth movement. At the end of the swing, the wrist rolls forward, accelerating the hammer to top speed just before it strikes.

HANDLES

The problem with wood-handle hammers is that they don't hold up well to a lot of tough nail pulling. I wrap several layers of electrician's tape around the top of the handle for reinforcement, but I still have to be careful when pulling nails. Steel-shank hammers such as the Estwing®, with the head and handle made of one-piece solid steel, are great for nail pulling. However, some carpenters who use steel-shank hammers complain of arm and shoulder fatigue without the wood handle to absorb the shock of driving nails.

CLAWS

Hammers come with either straight claws or curved claws, although the latter choice is reserved mainly for finish hammers. Older carpenters told me they preferred the curved-claw hammer because it easily pulled nails without marring wood. But every time I watched them pull a nail, they'd put a block of wood between the curved claws and their work.

FACES

Besides handle and shape of claws, the biggest difference between hammers is face treatment (see the photo on p. 42). I was thrilled when I got my first hammer with a milled face—until I missed and hit my finger. (There is a big difference between a mashed finger and a shredded finger.)

Milled-face hammers (also called serrated or waffle-face hammers) come with checkerboard grooves cut into the face. These grooves help to grab the head of a nail even when your aim might be off a little. When it's driving galvanized nails, the milled face doesn't become slick from the galvanized coating and slip off the nail like a standard smooth-face hammer. But be careful with your milled-face hammers around finish lumber: They can leave nasty hammer tracks.

Pick the Right Nail for the Job

In an effort to drive more nails faster, carpenters in the 1950s came up with the "gaswax" method of coating nails. First, they'd let a block of paraffin dissolve in about 4 gal. of gasoline in a bucket sitting in the sun. Then they'd splash a cup or so of the mixture into a 50-lb. box of nails. They'd shake the box to get all the nails wet. Then as the gasoline evaporated, the nails were left with a wax coat that made them a lot easier to drive. Looking back, I'm amazed that more job sites didn't go up in flames, to say nothing of gas-waxing as an environmentally unsound practice.

Thankfully, coated nails called sinkers made gas wax obsolete years ago. Sinkers have thinner shanks, and their coating makes them easier to drive than common nails. However, carpenters in other parts of the country use common nails exclusively. Using sinkers instead of common nails is a matter of preference, but building codes specify how many nails and what size must be used.

Codes vary depending on locale, so consult your building official. Codes require a given number of nails to hold framing members together at, for instance, stud-to-plate connections or floor-joist-to-rim-joist connections.

Many years ago, I was nailing door jambs in place with 8d finish nails that extended beyond the second piece of wood. An old craftsman on the job told me that if I used 6d nails instead, I'd save a mile of nails every year. A nail needs only to be long enough to grip the second piece of wood without going all the way through. No need to use a 16d nail to attach two 1×s.

Hardened nails are also available for attaching wood to masonry. It's best to drive these nails before concrete has fully cured. Masonry nails should never be driven with a regular hardened-steel hammer, which can cause chips of steel to break

The many faces of hammers. The hammer on the left has a standard smooth face for driving finish nails. The middle hammer has a face with a sandpaper texture that helps to keep the hammer on the nail, whereas the milled face on the right hammer grips the nail aggressively for framing.

Before milled-face hammers were available, we used to rub the smooth faces of our hammers on concrete to give them a slightly rough texture, which made them better at gripping nails. Dalluge now makes a finish hammer with a face that has been textured slightly to the consistency of 180-grit sandpaper. I am amazed by how much easier it is to drive a finish nail with this hammer rather than a smooth-face one.

WHAT TO CHOOSE

For framing my choice is a 22-oz. straight-claw hammer with a serrated face and an 18-in. wood handle. I like the balance and shape of this hammer, and I use the claws to pick up anything on a job site from a 2×4 to a heavy beam. I've also stuck those straight claws into roof sheathing to save me from sliding off an icy roof.

For a finish hammer I've always used a 16-oz. to 20-oz. head, straight-claw hammer with a smooth face and wood handle, although the slightly heavier Dalluge with the textured face is really appealing. The straight claws are easy to slip between a trimmer and a king stud when setting a door frame, and they even come in handy for tightening a wing nut.

off and fly through the air. Instead, I use a softer-headed hammer when I'm driving hardened nails.

Getting a Grip

Almost every carpenter has heard tales of flying steel and flying nails. Unfortunately, these stories are not just fairy tales with happy endings. Always wear approved eye protection. Never strike the hard steel of a hammer face against another hammer face. Small pieces of steel can break off and become dangerous projectiles. Flying nails can be just as dangerous. In the early 1950s, I hit a nail that flew and punctured my eyeball. Several stitches and several months later, I was back to work, wiser to say the least. It is not hard to see the moral of these stories. Eye protection isn't optional when driving nails.

Driving nails has more to do with rhythm, coordination, and timing rather than power and force. Your shoulder, elbow, forearm, and wrist should all work in a single fluid motion when driving large nails (see the illustration on p. 41). Smaller nails such as finish nails are driven mainly with the wrist and forearm.

Don't grab the handle with a tiring, white-knuckle grip. Instead, hold the handle near the end easily but firmly. Your thumb should wrap around the handle rather than resting on top, pointing toward the head. If the handle is slippery and difficult to hold, try roughing it up with a rasp or heavy-grit sandpaper. You can also smear a bit of wood pitch onto the handle to improve your grip. But go easy on the pitch, or you may end up sleeping with your hammer.

Place the nail on the wood, start it with a tap, and begin driving. A smooth, practiced swing will let you apply the maximum force with the least effort to drive the nail into the wood.

Nailing in a bent-over position, such as when assembling walls on a deck, is pretty easy if your aim is true. The weight of the hammer coming down reduces the amount of physical force you have to exert. The

TIP

If I have to drive a nail into fully cured concrete, I drill a small starter hole with a hammer drill. Without a starter hole, the nail may bend or may chip out a hole in the concrete.

For Nail Aprons, Simpler Is Usually Better

White bib overalls were a carpenter's trademark in the 1950s, and I miss them, with their pockets that held different nails, the pencil slot in the front, and that funky side pocket for my folding rule. There was even a loop for my hammer. But best of all, the overalls kept us warm in winter.

Cloth aprons came next, but they didn't hold enough nails. Webbed belts with leather nail bags sewn in front replaced the apron, but carpenters had trouble getting their hands into the nail bags while working bent over. Finally, someone came up with a wide leather belt holding removable bags, arguably the most popular nail-apron design today.

Carpenters like this style because they can position the bags wherever they like, and the wide belt helps to distribute the weight. I usually wear two bags, one for 8d nails and one for 16d nails, with smaller pouches sewn in for hanger nails, a measuring tape, and a small square. The latest design is the Cordura® nylon nail bag that is lightweight and hard to tear. The padded nylon belt is comfortable, but most of the nylon bags have more pockets than the average carpenter needs.

The first time I attached a pair of suspenders to my nail belt, my lower back breathed a huge sigh of relief. A loaded nail belt can weigh over 12 lb., and suspenders transfer most of that nail-belt weight to the shoulders.

same holds true when nailing plywood decking to floor joists.

Holding a framing hammer near the end of the handle when you're driving large nails into 2× stock maximizes the force you are applying to the nail. But less force is needed to sink smaller nails, so I choke up on the handle when driving 6d or 8d nails. It hardly takes a tap to drive an 8d into a green 2×4, and with less swing, choking up on the handle keeps the hammer head on target.

Grabbing the hammer handle more toward the middle also gives you greater control, so I also choke up on the hammer when driving finish nails. For most finish nails, I strike the nail using mostly wrist action to reduce the chances of missing the nail and marring the wood.

Continuous nailing. Fast and efficient nailing calls for nailing with one hand while feeding nails with the other (photo below left). First, get all the nails going in the same direction by pulling out upside-down nails and flipping them over (photo below right).

Don't Take Nails from Your Pouch One at a Time

Back before the days of collated coils and sticks of nails for pneumatic guns, carpenters would spend evenings bundling small bunches of nails with rubber bands to prepare for the next day's work. But then as now, I just pick up a handful of nails straight out of the box and hold them in a bundle with all the shafts parallel. Next I grab all the heads that are facing toward me with the fingertips of my other hand, pull them out of the bundle and turn them so that all heads are facing away from me (see the photo below right). Then I feed the nails out one at a time with my thumb and forefinger. As I drive one nail with my right hand, my left hand feeds out the next (see the photo below left). Eventually, a rhythm develops, and you'll be able to feed out nails as fast as you drive them.

Short nails, such as those used for roofing and drywall or for attaching framing anchors, are easiest to start by holding the nails between your first two fingers (see the top left photo on the facing page). I grab a handful of these nails and feed them out one at a time, but I don't worry about getting them all going the same way. With my thumb I flip each into position between my first and second fingers with my fingernails facing down. A good roofer nailing by hand this way can almost keep up with an air gun.

Roofing nails are held between fingers. Because of their short shanks and large heads, roofing nails can be set more easily by holding them between your first two fingers.

A vehicle for nailing floors. When a lot of subfloor needs to be nailed off, a nailing buggy, a scrap of plywood on wheels, can save your knees. The carpenter sits on it, pushing backward and nailing as he goes.

Framing with your feet. Placing the ball of your foot on the plate and your heel on the stud will keep the stud on the layout and keep it from bouncing off the plate during nailing.

When I first started working in the large housing tracts, there were acres of subfloor to be nailed. Without a good nailing rhythm, we'd still be there nailing floor. Some carpenters who didn't like nailing off flooring on their knees made themselves a nail buggy, which is a piece of plywood with wheels screwed to the bottom (see the top right photo at right). A tray nailed onto one side held the nails, and the carpenters would sit on the buggy and push themselves backward, nailing the subfloor to joists as they went. If you try a nail buggy, make sure you don't roll off the side of the building.

Hands and arms aren't the only body parts carpenters use when nailing a house together. Legs and feet also play an important part. When I'm building a wall flat on the deck, I often use my feet to manipulate the stud into place (see the bottom photo at right). Once the stud is on its layout, I place the ball of my foot on the plate and my heel on the stud to hold the framing members upright and to keep them from bouncing apart during the banging. A stud lying flat can be flipped on edge with your foot, and I occasionally put my feet on both sides of a stud and lift it with both heels to bring it up flush with the plate.

Bending Nails: To Pull or Not to Pull

Part of getting into the swing of nailing is making sure the face of the hammer strikes the nail squarely. If the hammer is tipped to one side, the force delivered to the nail isn't centered on the nail, and the nail will bend. I have worked with framers who bent nails and never bothered to pull them, reasoning that they would be hidden with sheathing or drywall. But I think bent nails leave a frame looking shoddy.

Every carpenter bends nails. However, most practiced carpenters know that when a nail does bend, they can usually tip the hammer in the opposite direction and hit the nail again to make it straighten back up.

A block adds leverage. Slipping a block of wood between your hammer and the wood decreases the amount of force that it takes to pull a nail. The block also protects the wood from damage from the hammer.

Prying out a nail sideways. To reduce stress on your hammer, grab the nail in the claws and push the handle to the side.

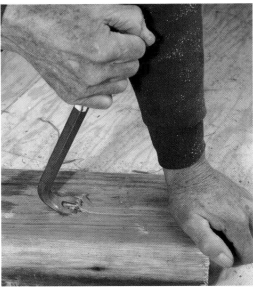

A cat's paw digs out buried nails. If a nail that's been punched home needs to come out, drive the claws of a cat's paw under the head of the nail to pry it out of the hole.

Bending nails is often the result of nailing in an awkward position or in a tight spot. So if you find yourself in such a position, use shorter, more direct strokes, and you'll bend fewer nails.

Sometimes, a nail will bend when it hits a knot or tough piece of wood. Generally, the harder the wood is, the harder it is to drive a nail without bending it. So when nailing into a hardwood, I hit the nail with softer, direct blows. Occasionally, I have sunk a nail through a knot by driving it through the center core, which is usually softer than the surrounding area.

If a nail bends too much to be straightened out by changing the angle of your hammer, it often can be straightened with the hammer claws and then driven home with softer blows. When a nail bends and you can't redrive it, pull it and try driving another nail in the same hole.

As I mentioned earlier, pulling nails can push a wood-handle hammer to its limit. I either slip a block of wood under the head to increase leverage (see the top left photo), or I slip the shank of the nail between the claws of my wood-handle hammer and push the hammer over to one side (see the top right photo on the facing page). So I release the hammer, hook the nail again, and push the hammer to the opposite side. I repeat the process until the nail is loose. Pulling nails in this fashion is effective but usually dents the wood pretty severely, so I don't try it with finish lumber.

If you are pulling a nail from an expensive piece of trim, put something softer than a block of wood between the hammer and the trim. A piece of heavy leather or rubber or folded-up cardboard works well.

There are times when a nail that has been driven home has to be pulled—maybe the head is in the way of a sawcut that you need to make. If I'm looking at a framing nail driven into 2× stock, I usually reach for my cat's paw (see the bottom photo on the facing page), which has short, hammerlike claws that come to a sharp point. I drive the points of the cat's paw into the wood just upgrain from the nail head. As the points dig into the wood, I angle the cat's paw toward the nail and drive the claws under the head. The nail can then be pried out enough to be pulled by a hammer.

When I need to remove a piece of trim while minimizing the damage, I use a fine nail set and drive the nails through the trim. Most nail sets will enlarge the hole slightly, but that is better than prying the trim off with a flat bar, a sure way to ruin a piece of trim. If I need to pull nails from a piece of trim that I've removed and want to save, I pull the finish nails from the backside of the piece using a pair of carpenter's nippers.

Toenails Work Best at the Proper Angle

Nails driven straight into end grain, like those through a wall plate and into a stud, have shear strength but little withdrawal resistance. On the other hand, toenails driven in at an angle will hold two pieces of wood together much more securely, especially when the nails in the same piece of wood are driven in at opposite angles.

If a toenail is started at the right spot and driven at the correct angle (about 60 degrees), half the length of the nail will be in each piece of wood. Most nails have a sharp point that lets you start them at the proper angle. But if you have trouble starting a toenail this way, you can tap the nail more or less straight into the wood and then pull it up to the correct angle before driving it home.

One place where I use toenails is for attaching cripples to load-bearing headers

Miss the Nail, and You'll Have to Cover Your Tracks

No discussion of nailing goofs would be complete without mention of what old-time carpenters called "Charlie Olsens." When you miss a nail and the hammer hits the wood, a telltale dent is always left. If the hammer is tipped a bit, the dent will look like the letter "C," and if the miss is full on, the dent will be Charlie's other initial, "O."

Like bent nails, Charlie Olsens are more likely to happen when nailing in an awkward position, and practice is the best way to keep Charlie away from the job. However, if you do happen to leave a dent in trim or baseboard where it will be seen, you'll have to do some fixing.

If the wood is to be painted, the holes can be filled with putty and sanded. Dents in stain-grade wood take a bit more work. A moist, cotton cloth placed on the dent and heated with a flatiron is often enough to expand and raise the dented wood. Once the wood has dried completely, give it a gentle sanding to blend the repair in with the surrounding wood.

Toenailing a cripple. Planting a foot behind the cripple keeps it from moving while toenails are driven in at an angle to secure the cripple to the header.

A toenail aligns crowned boards. To line up two boards that have to be nailed together but aren't parallel, first nail the ends, then sink a toenail through the highest part of the crown to bring the two boards into alignment.

A blunted nail won't split the wood. Flattening the point of a nail lets it break the fibers in a piece of wood as it's driven through instead of wedging the fibers apart and causing the wood to split.

(see the top left photo above). First, I start the nail in the cripple and then place it on the header about ¼ in. off the layout line. I back up the cripple with my foot and steady it with my hand while sinking the nails. The final hammer blow moves the cripple onto the layout line and draws it down tight to the header. Another toenail driven from the opposite side ties the cripple to the header. When driving the second toenail, try not to hit the cripple, which could loosen the cripple, knock it off its layout, and raise it off the header.

With practice, you'll be able to toenail framing members together just by holding them tightly in position with hand pressure. The key is making sure the angle of the nail is right. In these situations or when toenailing a stud to a sill plate, our code requires three 16d or four 8d nails. I use box nails because their thinner shank is not as likely to split the end of the 2×.

When toenailing cripples or driving nails near the end of a board, you'll sometimes need to avoid splitting wood with the nail. In these cases, I try dulling the point (see the bottom photo at left). Place the nail with the point upright and with the head on a hard surface. Tap the point several times with your hammer. The blunt point breaks the wood grain as the nail penetrates instead of spreading and splitting it.

I also drive toenails to straighten bowed framing members that have to be nailed together, such as two 2×s forming a header or a corner stud (see the top right photo on the facing page). This is one place where pneumatics will never replace the good old hammer. If one board is crowned so that it stands higher than the second board, I nail both ends together and then drive a toenail into the highest part of the crown, banging the nail down until the crowned board is flush with its partner. A couple of 16ds are then driven through the face of the header to keep the two boards aligned permanently.

Change Your Swing for Nailing in Tight Places

One way of nailing in tight places such as under a roof overhang or toenailing joists to plates is with short strokes: tap, tap, tap. But in most cases, I take a modified full swing, moving the hammer not only in an arc, but also up as it clears the overhang (or down as the hammer clears the joist) and approaches the nail.

These tight spots are prime candidates for bent nails and Charlie Olsens, and one way to help prevent those mistakes is by angling the nail toward you slightly before trying to drive it home. Also, if you're nailing in that last piece of gable siding over your head, get the nails started before climbing up the ladder.

There are also times when you'll need to drive a nail that's a full arm's length away, such as way over your head (see the bottom photo above). Starting the nail is the trick in these situations. Today, you can buy a hammer with a magnetic slot in the head where you can insert a nail and start it with your first swing. Another option I use is grabbing the head of my framing hammer

Starting a nail one-handed. When you need to drive a nail that's beyond the reach of two hands, wrap one hand around the head of the hammer and hold the head of the nail against the hammer head with your thumb and forefinger.

Remote control. With the nail held as shown, you can stretch up to that out-of-reach spot and start the nail by swinging your fist toward the wood. Once the nail is started, regrip the handle and drive the nail.

in the palm of my hand. I hold a nail with my thumb and forefinger, placing the head of the nail against the side of the hammer head (see the top photo above). Then, with a quick punching action, I can start the nail with one hand.

Larry Haun, author of The Very Efficient Carpenter *(The Taunton Press, Inc.) and* Habitat for Humanity How to Build a House *(The Taunton Press, Inc.), has been framing houses for more than 50 years. He lives in Coos Bay, Oregon.*

Anchoring Wood to a Steel I-Beam

■ BY JOHN SPIER

Modern floor plans are trending toward wide-open spaces. Despite advances in engineered-wood beams, there are times when something stronger is needed. Many carpenters shy away from steel because fastening lumber to steel can be tricky. Cutting a steel beam on site is even trickier. Sometimes, though, a steel I-beam is the best choice. Structural steel costs less than comparable LVLs, is strong, and is available from local suppliers. If you order it to the right size with fastener holes punched, your only challenge will be attaching the lumber.

Steel Has a Few Limitations

Although a piece of steel carries a larger load over a longer span with less depth than any other building material, steel has some disadvantages. First, it's very heavy. You need to make sure you can get it to where it

needs to go, either with humans or with a machine. Second, you won't find steel span charts in a codebook; steel usually needs to be sized by an engineer. Steel should be protected from moisture to prevent rust and deterioration. Also, a steel beam will fail much more quickly and catastrophically than an equivalent wood beam in a fire.

Top-Mount Joist Hangers Are the Most Practical

In the simplest usage, an I-beam rests in pockets cast in foundation walls, with floor joists on top of it. More often, though, wood is bolted to the web; then joists or rafters are attached to the wood with standard joist hangers. The drilling and bolting required by this method are so impractical that it's worth changing before construction. Another attachment method is to weld top-mount hangers to the I-beam. However,

The Fastest Way to Marry Wood to Steel

Bolt a 2×6 or 2×8 to the top flange of a steel I-beam and use top-mount joist hangers to support the floor framing. Fastening 1½-in. lag bolts through the bottom of the top flange is far faster than countersinking the heads of through-bolts from above.

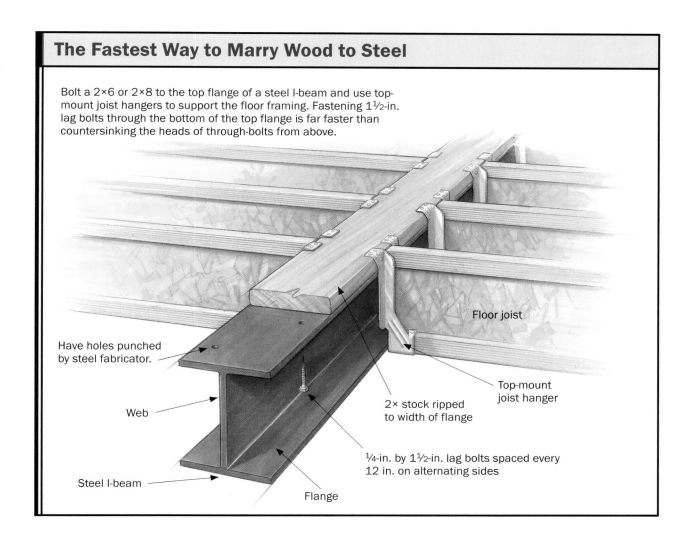

Have holes punched by steel fabricator.

Web

Steel I-beam

Flange

Floor joist

Top-mount joist hanger

2× stock ripped to width of flange

¼-in. by 1½-in. lag bolts spaced every 12 in. on alternating sides

---

this option is rare, not because it doesn't work but because most framing crews don't have a welder on site.

I think the best method is to bolt 2× lumber to the top flange of the I-beam, then nail top-mount joist hangers to the lumber (see the illustration above). I use ¼-in. by 1½-in. lag bolts through the bottom of the top flange. Because top-mount joist hangers won't resist twisting as well as face-mount hangers, I use strapping below or blocking between the joists. If uplift resistance is needed, I tack the lumber in place with a powder-actuated nailing tool, then through-bolt after the plywood subfloor is down, recessing the bolt heads into the plywood.

The steel I-beams most often seen on job sites are called W- (wide) and S- (standard)

A Waste of Time

Packing the web is a last resort. Unless there are compelling reasons (such as headroom issues) to attach wood this way, I avoid it. This is a very strong method, but it's too time-consuming to make it practical for everyday use.

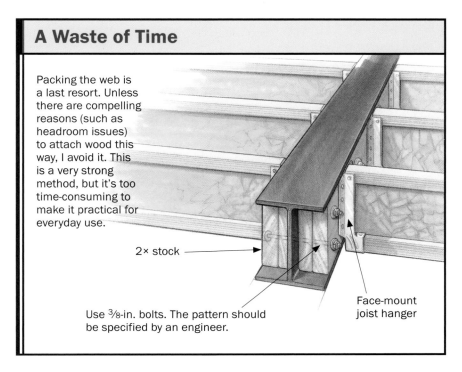

2× stock

Face-mount joist hanger

Use ⅜-in. bolts. The pattern should be specified by an engineer.

I-Beam Options

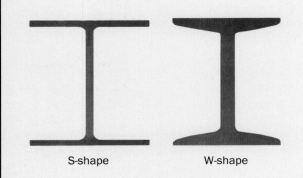

S-shape   W-shape

Most I-beams on job sites are W-shapes, or wide flange. Another common beam is S-shape, or standard, which has narrower flanges and a thicker web. S-shapes are generally taller than comparable W-shape beams. Narrower S-shape beams are better suited for recessing into wall framing, whereas W-shape beams are great where headroom is an issue.

shapes, depending on the width of the flange. Steel beams are designated by shape, depth, and weight. For example, a W8x35 beam is a W-shape about 8 in. deep and weighing 35 lb. per lin. ft.

Because steel is hard to cut and drill on site, your supplier needs the exact length along with sketches showing all hole locations and sizes. If you have to cut a beam, an acetylene torch is the easiest method, but metal-cutting blades in circular saws and reciprocating saws work, too.

Steel Needs Protection from the Weather

To prevent corrosion, most suppliers spray steel with a primer. It's worthwhile to make sure primer is applied because it keeps the steel clean and rust-free while you're working with it. Occasionally, galvanizing is specified for steel components. If it is, make sure all the cutting and drilling are done first, and remember that the holes need to be oversize by ⅛ in., or the bolts won't fit through after galvanizing is done.

Really big beams are best set in place with a crane, but lesser beams often can be placed with a lumber-delivery boom truck. A backhoe or excavator works if it can get close enough. I've found that beams as heavy as 600 lb. or so can be set safely with human power (four to six people), as long as they're not too high up. Use levers, rollers, winches, platforms, and plenty of caution. Finally, remember that wood shrinks and swells, but steel won't budge.

John Spier is a builder on Block Island, Rhode Island. His book, Building with Engineered Lumber, *is available from The Taunton Press.*

Common Engineering Problems

■ BY DAVID UTTERBACK

Over the past 22 years as a builder, building inspector, and lumber-industry representative, I've inspected a great deal of framing in all parts of the country. Terms and techniques vary from region to region, but mistakes don't. The same problems tend to show up over and over.

Here, I'll examine some of these problems from an engineering standpoint and look at what can be done to avoid them. All these situations are addressed in similar ways by each of the three major building codes. Before going further, I must emphasize that difficult framing problems often require complex engineered solutions. When the going gets tough, your best bet is to enlist the services of a good engineer. It's a lot cheaper than defending yourself in a lawsuit.

Bracing for bad news. Placing too big a pile of sheathing on an unbraced truss roof can lead to disaster. Amazingly in this case, the falling dominos were halted when a quick-thinking carpenter was able to brace the remaining upright trusses before the collapse reached the part of the roof where he was working.

Joist-Hanger Nails Are Not Meant for Installing Joist Hangers

Joist hangers are marvelous devices for supporting joists or beams that cannot rest directly atop vertical framing members. To get the most structural capacity from a joist hanger, you must use the correct hanger for the joist and place the right nail in every nail hole.

Many builders mistakenly assume that the right nails for every situation are the 1½-in. long "joist-hanger nails" sold by the manufacturer. In truth, these nails are intended for anchoring the sides of the hanger to a single joist without piercing the other side.

Joist-hanger nails have the same diameter, and therefore the same shear capacity, as 10d common nails, but their shorter length gives them less withdrawal resistance. For maximum strength, nothing smaller than 10d common nails (or 16d sinkers, which have the same diameter) should be used to attach a single joist hanger to a beam (see the illustrations below). To attach a double joist hanger to a beam, 16d commons should be used.

This fact does not mean that you can never use the short nails to support a joist hanger. But if you do, you must reduce the load. If joist-hanger nails are used instead of 10d commons to support a single hanger, you can use only 77 percent of the load value of that hanger. If they are used instead of 16d commons to support a double hanger, that load capacity drops to 64 percent. It's always wise to check with the hanger manufacturer if you are not sure what size nails to use. Some hangers have the required nail size stamped directly on the hanger.

Besides nails, you also need to understand the differences between hangers. Some hangers have little dog ears on the side of the hanger sticking out at 45-degree angles (I've seen framers bend the ears over to get them out of the way). These hangers require what is called double-shear nailing: Common nails are driven through these holes at an angle into the joist and on into the supporting beam or header, distributing the load through two points on each joist nail for greater strength. If you use this type of hanger, make sure it is nailed correctly.

Joist-Hanger Nails Are Too Short for Some Applications

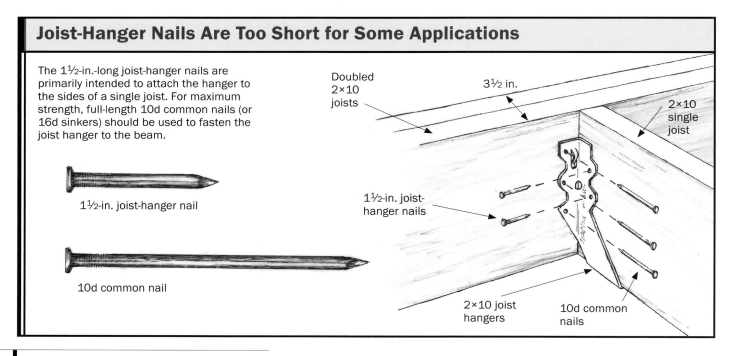

The 1½-in.-long joist-hanger nails are primarily intended to attach the hanger to the sides of a single joist. For maximum strength, full-length 10d common nails (or 16d sinkers) should be used to fasten the joist hanger to the beam.

1½-in. joist-hanger nail

10d common nail

Doubled 2×10 joists

3½ in.

2×10 single joist

1½-in. joist-hanger nails

2×10 joist hangers

10d common nails

Nonengineered Cantilevers

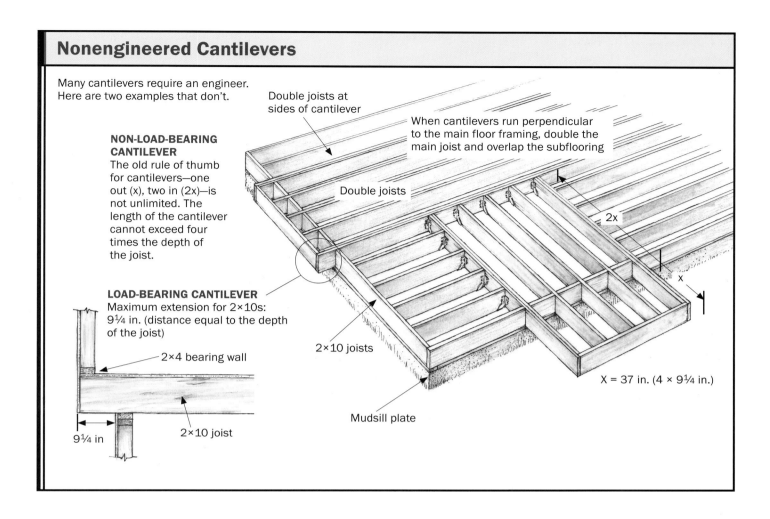

Many cantilevers require an engineer. Here are two examples that don't.

Double joists at sides of cantilever

When cantilevers run perpendicular to the main floor framing, double the main joist and overlap the subflooring

NON-LOAD-BEARING CANTILEVER
The old rule of thumb for cantilevers—one out (x), two in (2x)—is not unlimited. The length of the cantilever cannot exceed four times the depth of the joist.

Double joists

2x

LOAD-BEARING CANTILEVER
Maximum extension for 2×10s: 9¼ in. (distance equal to the depth of the joist)

2×10 joists

x

2×4 bearing wall

2×10 joist

9¼ in

X = 37 in. (4 × 9¼ in.)

Mudsill plate

Load-Bearing Cantilevers Need Careful Engineering

Many builders lay out cantilevers according to a simple rule of thumb: "One out, two in." This rule means that for whatever length the joists extend past their bearing point, they should run back in at least twice as far. Although technically correct, the rule applies to non-load-bearing applications only, and even then has its limits. In nonbearing applications, a joist may not cantilever more than four times its depth. Therefore, a 2×10 joist should cantilever no more than 37 in. (4 × 9¼ in.), regardless of its length (see the illustration above).

Non-load-bearing cantilevers can include sun decks and even bay windows (the cantilever supports only the weight of the window; any loads above are carried on a header set into the main wall). On the other hand, a zero-clearance fireplace with a two-story wood-frame chase would impose a significant bearing load on a cantilever.

Some simple load-bearing cantilevers can be built without paying an engineer. Because loads transfer through solid-sawn joists at 45-degree angles, codes allow load-bearing cantilevers that extend the same distance as the joists are wide. In other words, you could set a bearing wall on the end of 2×10 floor joists that are cantilevered 9¼ in. without risking a correction notice (see the detail illustration above).

Cantilevered joists that run perpendicular to the main floor joists may have another problem: If the connection between the two is not constructed properly, a teeter-totter effect could force the inside edges of the cantilevered joists upward, creating a hump in the floor.

To prevent this unpleasantness, the cantilevered joists should butt into a main joist that has been doubled to serve as a header. The connection between the cantilevered joists and the header should be securely constructed to prevent independent movement. As an added measure, subflooring should overlap the joint where the cantilevers and the main joist meet.

Bearing Walls Should Line Up with Their Supports

To transmit loads smoothly from roof to foundation, bearing walls must be stacked closely above one another. Where I-joists are involved, each bearing wall must sit directly over the top of its support because the web of an I-joist has little cross-sectional strength. Solid blocking or squash blocks also need to be installed according to the manufacturer's instructions to carry the load around the web and prevent the web from buckling.

Solid-sawn floor joists have more cross-sectional strength than I-joists, which allows you a little bit of leeway if you need to offset a bearing wall from its support. You can basically treat this situation the same as you would a load-bearing cantilever, meaning you could offset the bearing walls the same distance as the depth of the floor joist. If you had a 4½-in.-wide flange supporting 2×10 floor joists, you could set a bearing wall 9¼ in. to each side of the beam's edge and still meet code, giving you almost 2 ft. to play with (see the illustration below).

To prevent rotation of the joists, the codes also require full-depth solid blocking over beams or over bearing walls that support floor joists. As lateral loads, such as wind, are placed on the building, they're transferred into the floor diaphragm through the joists on their way to the foundation. By themselves, the nails that attach the plywood subfloor to the joists do not have the strength to resist these forces; if the floor joists are not blocked, they could actually roll over and end up lying flat.

Supporting Bearing Walls

When solid-sawn floor joists are used, bearing walls may be offset a distance equal to the depth of the joist. To keep the joists from rolling over, full-depth solid blocking is required between the joists where they rest on the bearing walls.

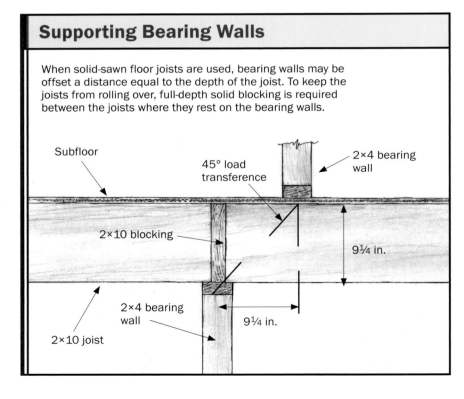

Subfloor

45° load transference

2×4 bearing wall

2×10 blocking

9¼ in.

2×4 bearing wall

2×10 joist

9¼ in.

Garage Walls and Cripple Walls Need Extra Bracing

Most regions of the country aren't threatened by earthquakes, but nearly every place is exposed to high winds. It is extremely important that walls be properly braced to resist these lateral loads, or the results could be catastrophic (see the photo on the facing page). Builders can generally rely on structural sheathing to brace walls, but that's not always enough.

Among the weakest points in a house frame are the narrow return walls on the sides of the garage door. Tall narrow walls are inherently difficult to brace properly against high lateral loads; this fact is why the Uniform Building Code (UBC) now

An otherwise well-built house. Because the tiny return walls on each end of the three-car garage could offer little resistance, lateral loads from an earthquake literally twisted this house off its foundation. The structural integrity of the rest of the house was largely unaffected by the earthquake.

requires a minimum 2-ft. 8-in. width for garage return walls. If you absolutely must squeeze in space for three cars, you can build a shear wall on site—by following a precise schedule for framing, nailing, and bolting—that will allow you to reduce this width to 24 in. or possibly even 16 in. (see the illustration at right).

Cripple walls (short kneewalls that run between the mudsills and the first-floor joists) are another weak link in the structural chain. Besides transferring vertical loads through to the foundation, these walls must also resist lateral loads. Cripple walls are effectively shear walls, and as such, they must be braced with structural panels and nailed 6 in. o.c. to provide the shear resistance necessary to support the structure above.

Another bracing point often overlooked is the connection between first-story and second-story walls. Most of the time, builders brace these walls independently of each other. Because lateral loads such as high wind can impose torque (turning or twisting energy) on a building, the upper story will move more than the lower story if the two aren't tied together.

Wider Is Better

Conventionally framed garage return walls must be wide enough to resist lateral loads imposed by high winds or ground movement. The Uniform Building Code minimum width of 2 ft. 8 in. can be reduced, however, if precise shear-wall schedules are followed. The following is one example of a 16-in. shear wall.

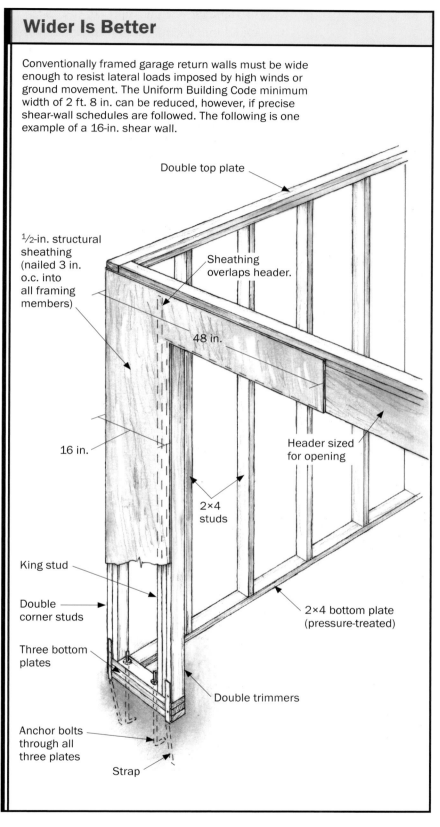

Double top plate

½-in. structural sheathing (nailed 3 in. o.c. into all framing members)

Sheathing overlaps header.

48 in.

16 in.

Header sized for opening

2×4 studs

King stud

Double corner studs

Three bottom plates

2×4 bottom plate (pressure-treated)

Double trimmers

Anchor bolts through all three plates

Strap

Fortunately, these walls can be tied together easily. One solution is to overlap the sheathing panels between floors (blocking the panel edges may be necessary in areas that are subject to high lateral loads). If sheathing is already in place, another solution is to tie the lower studs to the upper ones using metal straps specifically manufactured for that purpose.

Think Twice before Cutting Beams

It's easy to pull out a saw and cut off the top corner of a beam that must be kept beneath a roofline. But if too much cross section is removed, shear forces can cause the beam to split and eventually to fail. For solid-sawn beams, you should leave at least half the width of the beam above the supporting wall and confine the length of the tapered cut to no more than three times the original width of the beam. If you don't have room to leave this much cross section, your best bet is to lower the beam (set it in a pocket) or have a tapered beam engineered.

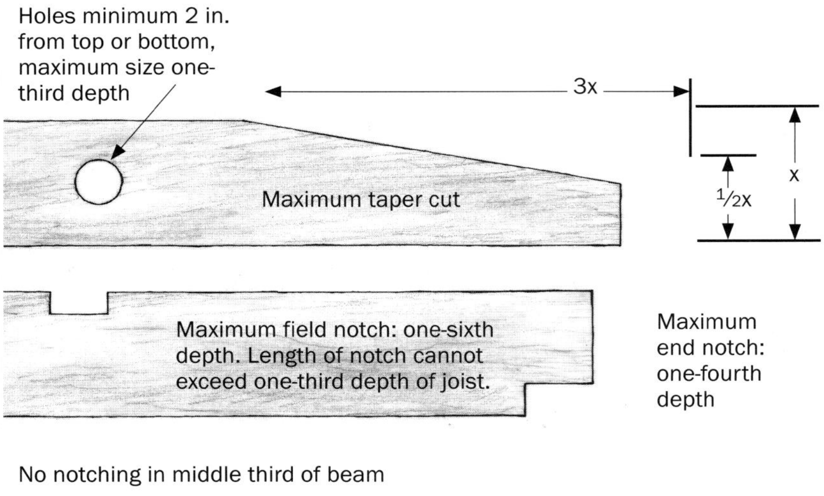

Solid-sawn beams may be notched one-quarter their depth at the ends and one-sixth their depth in the outer thirds of the span. Holes may be drilled in a beam from face to face, but never from edge to edge. The diameter of the hole may be as big as one-third the depth of the beam, but it must be at least 2 in. from the top or bottom edge (see the illustration below).

Rafter Ties Must Be near the Plates to Be Effective

Many builders confuse collar ties with rafter ties. Both are horizontal framing members that connect rafters, but that's where the similarities end. Collar ties (which are required by the Southern Building Code and no other) function to resist the pressures of wind uplift on a roof by holding the rafters together where they meet the ridge. As high up as they are, collar ties have no leverage to prevent the rafters and walls from spreading outward. That job is best done by the ceiling joists (see the illustration on the facing page).

If there are no ceiling joists or if the joists run perpendicular to the rafters, then the code requires rafter ties. Similar to a ceiling joist, a rafter tie is typically a 2×4 that runs parallel to the rafters, from outside wall to outside wall, and ties the rafters together as close to the top plate as possible. Rafter ties need to be installed every 4 ft. down the length of the roof.

Rafter ties do not have to be at ceiling height to be effective, but they must not be placed any higher than the lower third of the roof pitch. In other words, measure vertically from the outside wall's top plate to the bottom of the ridge and place the rafter ties within the lower third of that measurement. Once they get above that point, they lose their most effective leverage.

I've seen builders compound their mistakes when they try to use rafter ties as ceiling joists in semivaulted ceilings. For

Where to Cut or Drill Beams

Plumbers, electricians, and HVAC installers are as guilty as carpenters when it comes to carving—and weakening—beams. Here's a brief rundown on what the codes allow.

Holes minimum 2 in. from top or bottom, maximum size one-third depth

3x

Maximum taper cut

1/2x

x

Maximum field notch: one-sixth depth. Length of notch cannot exceed one-third depth of joist.

Maximum end notch: one-fourth depth

No notching in middle third of beam

The Wrong and the Right of Rafter Ties

To prevent roof loads from spreading the walls outward, rafter ties (or ceiling joists) must be in the lower third of the roof pitch. Collar ties are too high to keep walls from spreading and instead serve to resist uplift by holding the rafter together at the ridge.

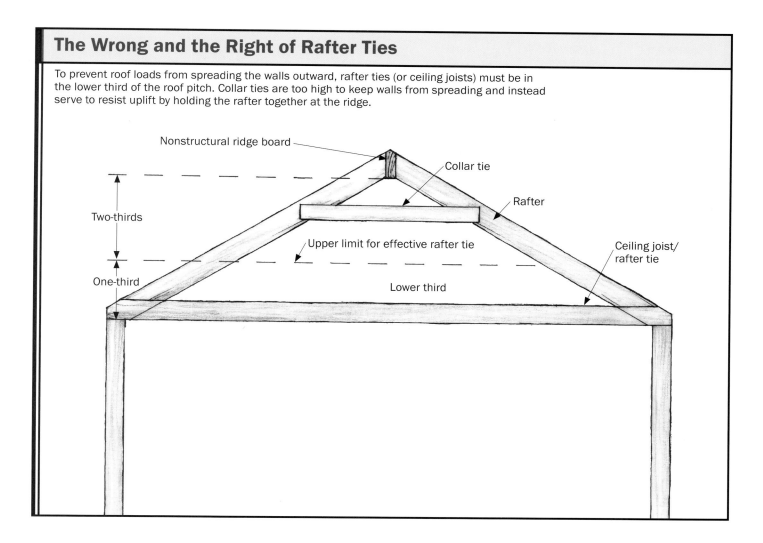

Nonstructural ridge board

Collar tie

Rafter

Two-thirds

Upper limit for effective rafter tie

Ceiling joist/ rafter tie

One-third

Lower third

maximum headroom or aesthetic balance, they place the rafter ties halfway up the roof pitch, near the center of the rafter span, where they're too high to be an effective tie. Applying the insulation and the drywall greatly increases the load on the rafters at their most critical point: midspan (what engineers call the maximum bending moment). This added load can cause the rafters to sag, pulling the ridge down and also pushing the exterior walls outward.

To avoid this problem, you'd need to engineer the rafters to carry the point load created by the additional weight being placed on them. You'd also need to design a ridge beam capable of supporting the roof load, just as you would if it were a cathedral ceiling, which essentially it is.

Trusses Require Precise Permanent Bracing

Any builder who's ever heard the words *domino effect* knows it's important to brace trusses as they are being erected. But not everybody understands what permanent bracing involves.

To ensure a stable, long-lasting roof, most truss systems require three types of permanent bracing: continuous lateral bracing of the top and the bottom chords and diagonal bracing at the end of the building, and in between if necessary (see the illustration on p. 60). The bracing for the top chord is typically satisfied when the roof sheathing is applied. The bottom-chord bracing

> *Any builder who's ever heard the words domino effect knows it's important to brace trusses as they are being erected.*

Permanent Truss Bracing

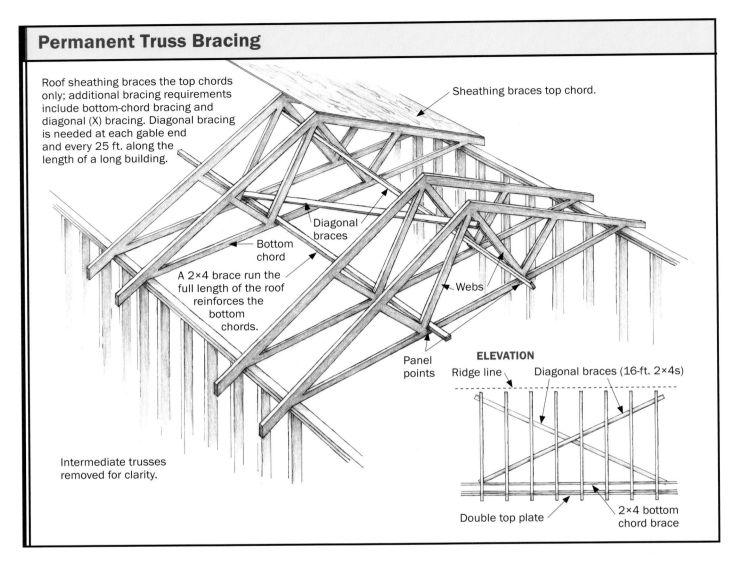

Roof sheathing braces the top chords only; additional bracing requirements include bottom-chord bracing and diagonal (X) bracing. Diagonal bracing is needed at each gable end and every 25 ft. along the length of a long building.

Sheathing braces top chord.

Diagonal braces

Bottom chord

A 2×4 brace run the full length of the roof reinforces the bottom chords.

Webs

Panel points

Intermediate trusses removed for clarity.

ELEVATION

Ridge line

Diagonal braces (16-ft. 2×4s)

Double top plate

2×4 bottom chord brace

is normally accomplished by placing a row of 2×4s on top of the bottom chord and then running them alongside a panel point (the point where the webs and the bottom chord meet) for the full length of the building. In a wide building, these bottom-chord braces should be roughly 10 ft. o.c.

The diagonal bracing—actually a form of X-bracing—is the one builders often get wrong or omit altogether. Diagonal bracing should be placed at each end of the building and every 25 ft. along the length of a long building. To prevent the domino effect, the first leg of the X is formed by a 16-ft. 2×4 running down at a 45-degree angle (or less) from the ridgeline of the gable-end truss to the bottom chord of the farthest reachable

inner truss. To take the hinge effect out of the connection between the gable wall and the gable truss, the other leg of the X runs from the top plate of the gable-end wall upward to the top chord of the same inner truss to which the first leg is attached. It's also important to make sure the braces are run alongside the webs of the intervening trusses and securely nailed to each truss.

In certain situations, it may also be necessary to brace long web members that are in compression to prevent them from buckling under load. If any web bracing is required, the proper procedure for it will be noted in the technical design sheet that comes from the manufacturer.

Loads Must Be Placed Carefully atop Trusses

After the trusses are placed and braced—and before the crane operator is allowed to leave—many builders lift pallet loads of sheathing panels onto the trusses for easy access by the framers. Although this practice may be convenient, it can greatly overload the trusses if not done carefully (see the photo on p. 53).

On steep roofs, the crew typically erects a temporary platform to hold sheathing. This platform usually consists of two legs resting on the trusses, plus some framework to support the sheathing horizontally (see the sidebar below). This arrangement puts most of the load on two trusses. The higher on the roof these loads are placed, the greater the stress factor. Severe bowing or total collapse of the trusses can result if the stress is too great.

The proper way to load pallets of sheathing on the roof is to place the load as low as possible on the trusses. The type of platform I just described is fine, but set the legs down onto the top plate of the exterior wall. This way, the wall—not the trusses—shoulders most of the burden (see the sidebar below).

__David Utterback__ is a former builder and a certified building inspector who conducts seminars on building codes and wood-frame construction.

Which Roof Would You Rather Sheathe?

In the left photo, the sheathing has been stacked too high on this steep roof and is seriously overloading several trusses. A strong gust of wind is all it may take to bring down this house of cards. In the right photo, even though it's more than a full pallet, the load is low on the roof, and the weight is bearing on the walls.

The Future of Framing Is Here

■ BY JOSEPH LSTIBUREK

Back in the 1970s, as a young engineering student studying energy efficiency, I wondered, "When the price of oil doubles, will the walls we're building now look smart or dumb?" The answer was obvious: They'll look dumb. That's when I started my quest for the future of walls.

Contrary to Hollywood's advice in *The Graduate*, the future is not plastics. The present is plastics. The future is wood (actually, it's cellulose, the stuff wood's made of), and the future is now. That's good news for the United States, because we're the Saudi Arabia of cellulose. Saudi Arabia has sand and oil; we've got dirt and cellulose. Oil is nonrenewable, but cellulose grows on trees.

Smarter strategies. This Colorado subdivision illustrates that some builders are using smarter framing strategies. The minimalist skeleton, which makes room for more insulation, is visible in the house in the foreground and in the top photo on the facing page. The insulating skin, visible on the house in the background, boosts the R-value. The nearly finished product (center house) looks normal, but its energy performance is superior.

Windows fall on 2-ft. layout

Stacked framing

Headers sized properly

2×6 studs 24 in. o.c.

Common Questions

Q. What about shear strength?
A. When sheathing with 1-in. foam, shear strength can come from strategically placed ½-in. OSB covered with ½-in. foam or from site-built shear panels (see p. 68).

Q. What about bouncy floors?
A. Yes, removing every third joist could make the floors more bouncy, but using thicker subfloor (1⅛-in. panels) will stiffen it back up.

Q. What about blocking for drywall?
A. Using drywall clips and floating the corners (leaving them unattached to the framing) are excellent ways to reduce drywall cracks.

Q. What about flimsy walls?
A. Half-inch drywall over studs 24 in. o.c. isn't all that flimsy (especially over dense-pack cellulose), but if you don't believe it, then use ⅝-in. drywall.

A. Wood is not a good insulator.

More often than not, the R-value of walls is assumed to be the same as the insulation in them. But this assumption doesn't consider all the wood framing connecting interior and exterior surfaces. Thermal bridging across framing members reduces overall R-value because a 2×4 or 2×6 is a poor insulator compared to fiberglass or cellulose. That's why eliminating unnecessary framing in exterior walls is so important.

By the way, if you're interested in knowing the actual R-value for a given wall assembly, you can use the handy calculator at the Oak Ridge National Labs website: www.ornl.gov/sci/roofs+walls/AWT/ InteractiveCalculators/rvalueinco.htm.

Most framers think structurally, not thermally. This photo was taken three miles down the road from the houses on pp. 62–63. These massive thermal bridges and hard-to-insulate cavities will make this a frigid wall. Because it's a bathroom, a mold problem is likely.

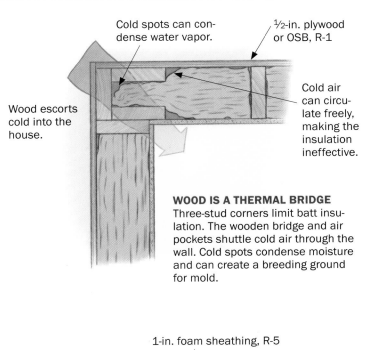

Cold spots can condense water vapor.

½-in. plywood or OSB, R-1

Wood escorts cold into the house.

Cold air can circulate freely, making the insulation ineffective.

WOOD IS A THERMAL BRIDGE
Three-stud corners limit batt insulation. The wooden bridge and air pockets shuttle cold air through the wall. Cold spots condense moisture and can create a breeding ground for mold.

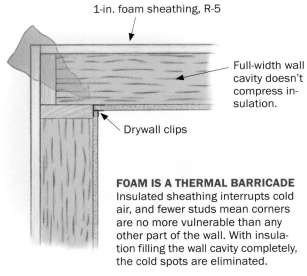

1-in. foam sheathing, R-5

Full-width wall cavity doesn't compress insulation.

Drywall clips

FOAM IS A THERMAL BARRICADE
Insulated sheathing interrupts cold air, and fewer studs mean corners are no more vulnerable than any other part of the wall. With insulation filling the wall cavity completely, the cold spots are eliminated.

The future lies in better wood products and better use of those wood products. OSB, engineered beams, and I-joists are already common products; in the future, we're going to get a lot more of these types of products.

To use all this "engineered cellulose" simply and elegantly, we need to convince hundreds of thousands of builders that the way they're building now no longer makes sense. Welcome to my world.

Smarter Walls Are Being Built Today

Extraneous studs, headers, and plywood don't boost structural integrity as much as they sabotage energy performance. For 30 years, engineers have been trying to convince us that the way we frame houses is inefficient; there's too much redundancy

A. They all make sense, but some give more bang for the buck.

You don't have to use all these details, but a couple of them will save you a bundle. Rather than switching all at once, start with the most efficient upgrades, then phase in new details after each is incorporated into your standard operating procedure. Cost savings are based on a 2,000-sq.-ft. house (see case study on pp. 66–67).

PHASE 1: DESIGN IN 2-FT. MODULES

The best thing you can do is to switch from 2×4 studs at 16-in. spacing to 2×6 studs at 24-in. spacing. Stack the floor, wall, and roof framing, and place windows and doors on the stud layout. Next, replace plywood or OSB wall sheathing and housewrap with at least 1 in. of rigid-foam sheathing. These steps will save you significant money and labor, and they'll boost R-value by 50 percent. Walls framed on the deck will also be much lighter and easier to stand up. Cost saving: $500.

PHASE 2: ELIMINATE COLD SPOTS

Structural headers aren't needed in non-load-bearing situations; size them properly in bearing situations. Corners and wall blocks make more cold pockets in a standard-frame wall. Use two-stud corners, and eliminate blocks to keep insulation consistent. Drywall can be floated at the corners or fastened with drywall clips. Cost saving: $135.

PHASE 3: FINE-TUNE THE SAVINGS

Use header hangers rather than jack studs at door and window openings. If cripples under windows are less than 24 in. tall, eliminate them altogether. This saves labor and materials, but may make trim installation more difficult. Eliminating one of the top plates is a final material-saving upgrade, although until precut studs are available at 94 in., this may complicate drywall installation. Cost saving: $120.

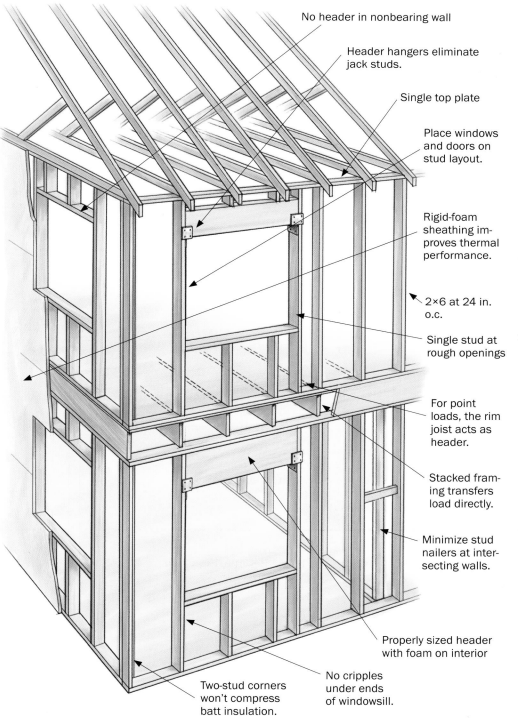

No header in nonbearing wall

Header hangers eliminate jack studs.

Single top plate

Place windows and doors on stud layout.

Rigid-foam sheathing improves thermal performance.

2×6 at 24 in. o.c.

Single stud at rough openings

For point loads, the rim joist acts as header.

Stacked framing transfers load directly.

Minimize stud nailers at intersecting walls.

Properly sized header with foam on interior

No cripples under ends of windowsill.

Two-stud corners won't compress batt insulation.

TIP

Because we can make sheathing, beams, joists, and rafters with small trees that are chipped up, we really don't need to cut down old-growth forests in the mountains to get the wood we need. We can grow trees on flat land, such as Ohio and Indiana. But cellulose also can be extracted from fast-growing plants rather than from trees.

even for them. But with houses and energy costing more than ever, it's time to listen.

As part of the U.S. Department of Energy's Building America program, our team focuses on the future of housing. Our target is an affordable "net-zero" house (one that produces as much energy as it consumes) built by production builders at no extra cost. Our target date is 2020, but I think that we can do it sooner.

To accomplish the goal of an affordable net-zero house, we have focused mostly on the enclosure. The enclosure of the future will be a lot like today's best enclosures, which use foam sheathings, housewraps, and spray insulations. But the materials of the

future will be smarter (more on that later), and framing redundancies will be gone.

The easy part of our job is figuring out how builders should be framing houses (see the sidebar on p. 65). Thirty years ago, the NAHB Research Center developed optimum-value engineering (OVE) to cut the cost of houses by omitting unnecessary lumber. OVE framing increases joist, stud, and rafter spacing to 24 in.; places doors and windows on stud layout; and demands that framing members be lined up (or stacked) for direct load transfer. Coupled with better insulation detailing, those same smart-framing strategies also can reduce the cost of heating and cooling houses.

Q. How can smart framing save money?

A. Fewer pieces go together faster, make less work for everyone, and leave more room for insulation.

A case study of two identical 2,000-sq.-ft. houses designed for a Centex Homes subdivision in Minnesota illustrates the magnitude of savings a single house can achieve. A comparison of wall elevations shows why one is cheaper to build, cheaper to heat and cool, and more polite toward

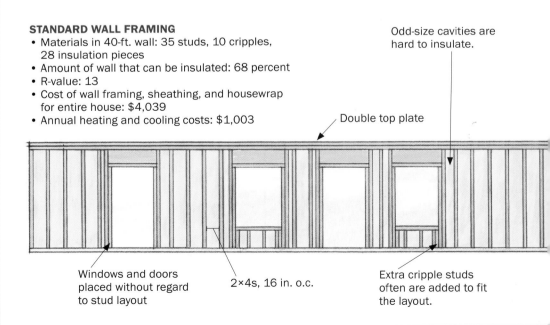

STANDARD WALL FRAMING
- Materials in 40-ft. wall: 35 studs, 10 cripples, 28 insulation pieces
- Amount of wall that can be insulated: 68 percent
- R-value: 13
- Cost of wall framing, sheathing, and housewrap for entire house: $4,039
- Annual heating and cooling costs: $1,003

Odd-size cavities are hard to insulate.

Double top plate

Windows and doors placed without regard to stud layout

2×4s, 16 in. o.c.

Extra cripple studs often are added to fit the layout.

Stack Framing Simplifies Load Paths

Lining up framing members directly on top of each other shouldn't be a big deal, but apparently it is because many builders don't do it. Stack framing makes everything simpler. Connections for high-wind, seismic, and high-snow-load areas are easier to detail, and mechanicals are easier to run when floor framing is spaced farther apart. You have fewer holes to drill and more room to work. Old-school builders may argue that framing on 24-in. centers makes bouncy floors, but if you glue and screw thicker sheathing, you can have a squeak- and bounce-free floor. The extra cost of thicker sheathing is offset by the lower cost of floor framing. Unfortunately, stack framing requires planning. Therein lies the problem.

Design Houses to Use Materials Efficiently

Because many materials come in 8-ft. sheets, we should account for that fact in our basic dimensions. We also should slide doors and windows to the nearest stud. As a hypothetical exercise, let's design two sheds out of OSB and wood studs. One is 8 ft. by 8 ft., and the other is 7 ft. by 7 ft. The materials list and the total cost of materials are the same for both. To figure out the cost per square foot for the 8-ft. shed, divide by 64.

environmental issues (such as greenhouse-gas emissions, resource conservation, and landfill congestion). Similar cost and resource efficiency also has been demonstrated on building sites in hot and mixed climates.

SMART WALL FRAMING
- Materials in 40-ft. wall: 21 studs, 2 cripples, 20 insulation pieces
- Amount of wall that can be insulated: 75 percent
- R-value: 24 (R-19 fiberglass batts, plus R-5 foam sheathing)
- Cost of wall framing and sheathing for entire house: $1,927
- Annual heating and cooling costs: $710

Single top plate

Doors and windows land on stud layout to minimize odd-size cavities.

2×6s, 24 in. o.c,

A. BUILD A SHEAR PANEL TO SLIP BETWEEN THE STUDS.

Insulating sheathing is an attractive alternative to OSB, but a major drawback is the lack of shear strength in a foam panel. One way to gain shear strength is to install ½-in. plywood or OSB at critical locations of a house, and then skin over it with ½-in. foam sheathing.

A better solution is a shear panel that fits into the wall framing, leaving the exterior foam intact. Leave one stud out and insert the 46½-in. panel. Built with readily available building materials for around $100, the panel is secured into the wall with nails and framing-connector plates and bolted continuously from the top plate to the foundation anchor bolts. This panel (developed and tested by Building Science Corp. and the Army Corps of Engineers) is engineered for site-built applications, but an engineer should specify where and how many to use.

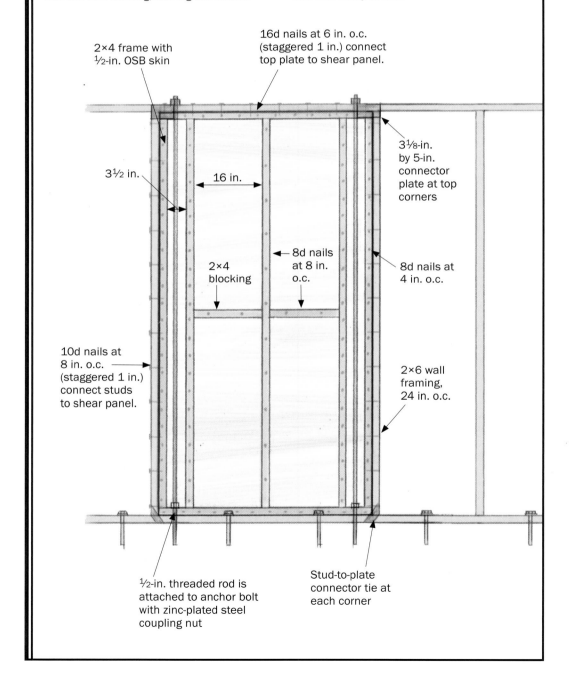

2×4 frame with ½-in. OSB skin

16d nails at 6 in. o.c. (staggered 1 in.) connect top plate to shear panel.

3½ in.

16 in.

3⅛-in. by 5-in. connector plate at top corners

8d nails at 8 in. o.c.

2×4 blocking

8d nails at 4 in. o.c.

10d nails at 8 in. o.c. (staggered 1 in.) connect studs to shear panel.

2×6 wall framing, 24 in. o.c.

½-in. threaded rod is attached to anchor bolt with zinc-plated steel coupling nut

Stud-to-plate connector tie at each corner

The cost per square foot for the 7-ft. shed is 25-percent higher because we now divide by 49. Which one is faster to build? And which one needs a Dumpster? A 23-ft. 6-in. size makes no sense to anyone except town planners, architects, and designers. If carpet comes in 12-ft.-wide rolls, it is dumb to have a bedroom 12 ft. 4 in. wide.

When Wood Moves, Drywall Cracks

In the words of renowned Danish woodworker and furniture maker Tage Frid, "Vood moves." Drywall doesn't like to move. It prefers to crack. The more you attach drywall to wood, the more drywall cracks you'll have, unless you let the drywall bend.

Remember drywall cracks caused by truss uplift? The solution was floating the corners: Let the wood move and the drywall bend. The same theory reduces drywall cracks at wall intersections and saves a bundle of studs. But don't just take my word for it. Here's proof: When we used smart framing with floated corners on a Building America subdivision with a production builder in Chicago, we reduced drywall cracks by over 50 percent. Because this builder frames 1,000 homes a year, his savings translate to about $500,000* per year on service calls.

Shear Strength Is a Big Deal

For plywood or OSB to provide shear strength, nails must be far enough from the edge of the panel that they don't tear the panel when under stress. With a double top plate, the panel can sit flush with the bottom plate and still have lots of "meat" to nail into at the top. Not so with a single top plate on a typical 8-ft. 1-in. wall frame. In fact, it just doesn't work.

The traditional solution is diagonal bracing, either metal straps nailed to the face of the wall frame or a 1×4 let in to the wall studs. Another solution is a commercially available inset shear panel, popular on the West Coast because of tremendous seismic activity. None of the shear-panel manufacturers we approached was interested in modifying a proprietary system for smart framing, but the Army Corps of Engineers was. Together, we developed an inset shear panel for 2×6 24-in. o.c. framing (see the sidebar on the facing page). This panel is available commercially (www.tamlyn.com), but the design and engineering works for site-built applications, too.

What Does the Future Hold?

In the future, building materials will work a lot harder. Foam sheathings will pass water vapor selectively if a wall gets wet. Housewraps will change characteristics depending on orientation, season, and climate. Ballistic housewraps will protect houses from projectiles in seismic and hurricane areas.

But smarter materials can't achieve their potential without smart building. Why aren't more houses built smarter? Because it's different. What we have is an inefficient framing system that we are all doing incredibly efficiently. We need to refocus on a more efficient system. The transition can be in phases to reduce the learning curve, but it still takes about 10 houses for a framing crew to execute smoothly. If you want to change, you will; but with any building boom, everything sells so quickly that there's no incentive to slow down your system. When the boom fades, change will take place. Of course, oil at $100 a barrel helps, too.

Prices are from 2005.

Joseph Lstiburek, *Ph.D., P. Eng., is the cofounder of Building Science Corp. in Westford, Massachusetts.*

Sources

U.S. Department of Energy's Building America program
www.buildingamerica.gov

EEBA Builder Guides
www.eeba.org

NAHB Research Center
www.toolbase.org
(search "OVE")

www.buildingscience.com (search "case study" and "shear panel")

Engineered Lumber

■ BY SCOTT GIBSON

Engineered structural beams made their North American debut in 1934 in Peshtigo, Wisconsin, just north of Green Bay. Glue-laminated timber (or glulam) had been widely used in Europe for decades, but when Max Hanisch, a German immigrant, suggested glulam for the new gymnasium at Peshtigo High School, building traditionalists balked. The Wisconsin State Industrial Commission demanded steel reinforcements, but Hanisch prevailed. He and his partners supplied arched glulam beams that are still in service nearly 70 years later.

That was then. Now manufacturers combine relatively small pieces of wood with adhesives to form a variety of structural members that can be three times as strong as the sawn timbers they replace (see the chart on p. 74). Glulam was followed in 1970 by laminated veneer lumber. Later, MacMillan Bloedel, the Canadian forest-products giant, invented two other engineered framing materials: parallel strand lumber, sold as Parallam®, and laminated strand lumber (LSL), sold as TimberStrand®. On the horizon are structural members that combine wood fiber with more exotic materials such as Kevlar® or carbon filament for even better performance (see the sidebar on p. 75).

Despite a Higher Cost, Engineered Beams Are in Demand

Builders and manufacturers both are reaping the rewards of engineered beams and headers. Engineered lumber supports much heavier loads than sawn timbers of the same size. Unlike sawn lumber, engineered lumber shows little if any shrinkage (reducing problems like sagging floors and cracked drywall) and has consistent design values. The downside: Engineered beams may cost three times as much as Douglas-fir dimensional lumber.

For manufacturers, engineered lumber represents a market opportunity just as supplies of tight-grained, old-growth timber dwindle and complaints about inferior fast-growth timber increase. Engineered wood turns spindly second- and third-growth timber—the very stuff many builders have learned to hate—into reliable framing material. Although not all green builders embrace the trend (see the sidebar on p. 75), high-tech manufacturing and materials make it possible to conserve timber by using species that are otherwise unsuitable for framing.

Continued on p. 74.

Exposed Parallam. A Connecticut builder liked the look of finished Parallam (see p. 75) enough to leave the central girder and ceiling joists exposed in his dining room.

A Primer on Common Engineered Lumber

LAMINATED STRAND LUMBER
Resembling oriented strand board, LSL (sold as TimberStrand) is glued-together strips of wood fiber, but its strands are shorter than parallel strand lumber, and not parallel.
- The weakest of the engineered materials, but usually the least expensive and the best value if it meets structural demands.
- New, stiffer version of LSL being introduced to compete with LVL.
- Commonly used for short headers where shear forces are a concern.

GLUE-LAMINATED TIMBER
Glulam is stacked, finger-jointed layers of standard lumber. Varying the grade of lumber used and the placement of different grades within the glulam tweaks performance.
- The only heavy structural framing member that can be manufactured in curves and arches.
- Can be pressure-treated.
- Can have built-in camber, an arch that combats deflection.
- A traditional choice for exposed applications.

LAMINATED VENEER LUMBER

Made from thin layers of wood glued together, LVL resembles plywood, except that the laminations all run parallel rather than perpendicular to each other.

- Not meant to be exposed to view or to the weather.
- Comes in thinner dimensions than other engineered lumber and is easier to handle by a small crew—no crane necessary.
- Stronger than sawn lumber or LSL, at a relatively low price.
- Commonly used as header material.

PARALLEL STRAND LUMBER

Sold as Parallam, parallel strand lumber is packed with long strips of wood fiber that are parallel to each other and saturated with adhesives.

- Best combination of strength and stiffness.
- Can be pressure-treated.
- Can come visually graded and finished for exposed framing.

Engineered Lumber versus Sawn Lumber

Given the same dimensions, engineered-wood beams are stronger than sawn lumber. The bar graph compares natural and engineered lumber for strength (load-bearing capacity) and stiffness (resistance to deflection under load) by assigning a value of 1 to No. 1 spruce-pine-fir (SPF). For example, laminated strand lumber is no stiffer than SPF, but it's nearly twice as strong. Compared with the same piece of SPF, parallel strand lumber (Parallam) is 50-percent stiffer and can carry three times the load. Common applications for engineered lumber include beams and headers that are unusually long, that carry an unusually heavy load, or whose depth is constrained by unusual height limitations.

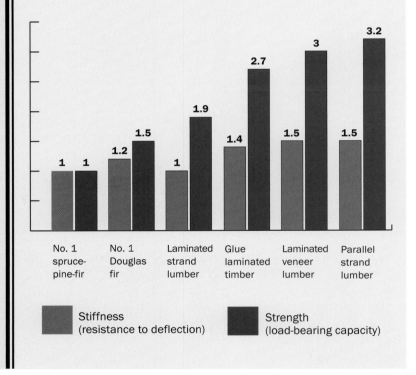

No. 1 spruce-pine-fir — 1, 1
No. 1 Douglas fir — 1.2, 1.5
Laminated strand lumber — 1, 1.9
Glue laminated timber — 1.4, 2.7
Laminated veneer lumber — 1.5, 3
Parallel strand lumber — 1.5, 3.2

■ Stiffness (resistance to deflection)   ■ Strength (load-bearing capacity)

A variety of wood species can go into engineered framing lumber. But because it's manufactured with specific design values in mind, the particular species doesn't matter much to the builder. Dan Harris, director of a Trus Joist® training center, compares it to buying Kentucky Fried Chicken℠: "It's like the Colonel. You're buying that piece of chicken, and it better taste the same every time. All the manufacturers test their stuff. Regardless of the species, the strength is there."

Glulam: High Strength in a Decorative Package

Glulam is stacks of dimensional lumber glued together like a wood sandwich. Where boards of one lamination meet end to end, manufacturers use finger joints for added strength. In the eastern United States, glulam usually is made of southern yellow pine, which is stiff, strong, and plentiful. In the West, Douglas fir and larch often are used, but other species also can be ordered. Glulam is graded for appearance and strength, and it can be pressure-treated for exterior use.

More than three dozen forest-product companies make glulam. (For manufacturers, see the websites of the American Institute of Timber Construction at www.aitc-glulam.org and the Engineered Wood Association at www.apawood.org). Glulam can be ordered in curved shapes to support a serpentine deck or in arches for ceilings. Stock widths run from 3⅛ in. to 6¾ in., including both 3½ in. and 5½ in. to make them compatible with 2×4 and 2×6 framing. Glulam also can be custom manufactured to just about any width and height.

In a simple span—that is, when a glulam spans an opening with no intermediate supports—it gets much of its strength from the bottom layer, or tension lam. Here, the Engineered Wood Association recommends what the industry calls an "unbalanced" beam. Unbalanced beams use laminations with different bending stresses for the compression (top) and tension (bottom) layers. Unbalanced glulam beams have a definite top. Where either the top or bottom may be stressed in tension, as in a cantilever, manufacturers recommend balanced beams, which are symmetrical in lumber quality.

Glulam is made of visually graded or, less frequently, mechanically stress-tested stock. By carefully selecting individual laminations, manufacturers can tweak design

Continued on p. 78

Builders Praise Performance but Debate Environmental Benefits

Rhode Island builder John Spier needed a dozen 20-ft. timbers to carry the second floor of the house he was building. Salvaged timber, although available, was expensive, and new timber might warp and shrink too much. His solution? Glulam.

Spier has used plenty of engineered-wood components. "They all have different strengths and weaknesses," he says. "Glulams are expensive and take quite a while to order, but they can be finished nicely (see the photo below). So for anything exposed, a glulam is the obvious choice. LVLs are built up, and I can get them in nice long lengths. They're very strong, and the companies will do the engineering work for me. A Parallam is kind of an ugly thing, but I can get it treated for use on exterior porch applications. It's very heavy but quickly available in my area, and it comes in big sizes."

Hank Fotter, a builder in Litchfield, Connecticut, disagrees with Spier about the looks of Parallam. He chose Parallam for an exposed use in his own dining room because he liked the way it looked (see the photo on p. 71). "I'm a convert," he says of engineered beams. "It's obvious to everyone that lumber is horrible coming off the truck." Fotter adds that its ability to eliminate drywall cracks and similar problems makes engineered lumber easy to sell to his customers.

But not everyone is on the bandwagon. Leo Ojala, a builder in western Massachusetts, uses as little engineered wood as possible. Past employment with a manufacturer of stress-skin panels left him with a chemical sensitivity to resins and adhesives used in engineered-wood products. He also worries that offcuts and job-site waste present serious disposal problems.

Although he's impressed with engineered lumber's strength and the fact that these products conserve diminishing natural resources, Ojala isn't swayed. "Is it a good use of forest products? Maybe, from one point of view. But if we're loading them up with glue and chemicals that are going to be omnipresent, are we doing a good thing? I don't know."

Glulam is a traditional choice for exposed framing. Rounded edges and a nice finish leave glulam looking good enough to use as an architectural feature.

The Engineers' Job Site: Inventing a Better Glulam

Current engineered beams are basically combinations of wood and glue, but more advanced products are on the way. One is a glulam supplemented with a layer of fiber-reinforced polymer.

Habib Dagher, director of the Advanced Engineered Wood Composites Center at the University of Maine, says reinforced beams offer greater span ratings and use less wood fiber than conventional glulam beams. Dagher and his colleagues are among a number of researchers around the country who have been working on these hybrids. The beams have been used on several bridge projects, and Dagher expects the know-how eventually will find its way into the residential-construction market.

Most glulam failures, he says, occur at or near the bottom lamination—the part of the beam under tension—either because a finger joint ruptures or because a defect in the wood weakens it. Replacing that lamination with a fiber-reinforced polymer layer sharply reduces the problem. The research center has experimented with Kevlar and carbon fibers and a variety of resins, epoxies, and thermoplastics.

If only 1.5 percent of the beam's depth is replaced with a polymer layer, the beam's strength roughly doubles. Stiffness goes up by another 15 percent to 20 percent. Consequently, a smaller beam can carry the same load as a regular glulam. Because fiber reinforcement is expensive, however, hybrid glulam starts to become cost-effective only at longer spans. Still, Dagher's work points to a new generation of engineered products that will be more sophisticated than a simple combination of wood fiber and glue.

Fresh from the lab and headed for market. This new glulam looks different because it's made of variable-size waste hardwoods instead of stock 2×s.

Beyond the breaking point. A polymer-reinforced glulam gets the ultimate stress test at the Advanced Engineered Wood Composites Center in Orono, Maine. The thin, dark polymer layer near the bottom roughly doubles the beam's overall strength.

strengths to suit different structural loads. A stock glulam will deflect sooner than either PSL or LVL. However, beefier versions, such as the Power Beam® from Anthony Forest Products (800-856-2372; www. anthonyforest.com), have higher design values to compete directly with other types of engineered beams.

LVL: A Radical Departure for the Forest-Products Industry

Trus Joist wasn't thinking beam or header when it developed laminated veneer lumber in 1970. Originally, Trus Joist designed its trademarked Microllam® as flange material for the company's wood I-joists. At the time, LVL was expensive, and the company didn't think it had much potential as a stand-alone product, Trus Joist's Harris says. Later, the company paired LVL beams and I-joists to give builders a fully engineered floor system, and LVL took off. LVL is manufactured by a dozen companies now, and roughly half of it goes into I-joists.

LVL is a lot like plywood. Thin veneers of southern yellow pine, Douglas fir, or other species are glued together to form a billet that can be cut to finished lengths and widths. Unlike plywood, though, the grain of the veneer runs in the same direction. Like other engineered-wood products, LVL doesn't shrink appreciably because the wood used to make it is dry when it's manufactured. LVL beams are straight, are essentially defect-free, and have consistent design values—all the qualities that dimensional framing material seems to lack.

Stock LVL comes in a variety of thicknesses and heights. The standard thickness is 1¾ in., meaning that two LVL components must be joined in the field to create a single timber compatible with 2×4 framing. That makes them easier to handle but adds to labor costs. Other widths, from 3½ in. to 5½ in., are available. Depths range to 20 in.

Most LVL has a bending strength slightly higher than that of standard glulam. But unlike glulam, LVL is available only as straight timbers and is not designed for exposed or exterior use. On the job site, uncoated LVL is susceptible to cupping when it gets wet.

Strand Technology Produces Two New Kinds of Engineered Beams

By the late 1980s, MacMillan Bloedel was becoming worried about supplies of softwood veneers that could be turned into plywood and LVL. It tinkered with ways of turning strands of wood fiber into lumber and invented two new types of engineered beams: parallel strand lumber, marketed as Parallam, and laminated strand lumber, which is sold as TimberStrand. Both still are protected by patents.

Although different in design and intended use, both of these structural components are manufactured by shredding trees into long strands and gluing the strands into larger structural components. Parallam starts with strands 2 ft. to 8 ft. long, arranged with wood fibers parallel to each other. They are bonded with an adhesive and run through a microwave to form beams up to 7 in. wide, 18 in. deep, and 66 ft. long. Finished beams have a stiffness similar to that of LVL and high-strength glulam. Unlike LVL, however, Parallam can be pressure-treated with wood preservative for use outside. Untreated versions swell if they become wet.

TimberStrand, Harris says, originally was designed not for structural framing but rather as a replacement for pine molding and millwork. But when door and window manufacturers didn't nibble, MacMillan Bloedel and Trus Joist, which formed a joint venture in 1991, looked for other uses. Manufactured much like OSB, TimberStrand is made from strands about 1 ft. long glued together to form sheets up to 8 ft. wide, 5½ in. thick, and 48 ft. long. It then is milled into timbers in widths of 1¾ in. and 3½ in. and in depths of up to 18 in. Laminated strand lumber is not as strong (or as expensive) as parallel strand lumber.

Choosing an Engineered Beam: Price, Availability, and Design

Glulam, LVL, and PSL all can be manufactured to handle heavy loads and long spans, says Steve Zylkowski of the Engineered Wood Association. As a result, selecting a particular brand or type of beam probably hinges less on strength than it does on price, appearance (if the beam will be visible when the house is finished), and ease of installation.

Builders see a hefty difference in price between dimensional and engineered lumber. TimberStrand and LVL cost a lot less than Parallam or glulam but still more than sawn lumber. Parallam and glulam, which always cost the most, are the only options for high design values and exposed use, but their prices and availability vary too widely by region to offer a general rule for choosing between them.

When an engineered structural component will be visible in the finished structure, architectural-grade glulam is the traditional choice. But Parallam made to be left exposed also is available. Choosing between Parallam and glulam might boil down to local availability, price, and the personal taste of the homeowner, engineer, or architect who writes the specifications for the job.

Job-site handling is a factor as well. Laminated veneer lumber may never win a beauty contest, but it typically arrives on site in 1¾-in. wide pieces that are nailed together to form larger-capacity beams. That means smaller crews can handle LVL beams. Parallam and glulam, however, arrive as single components, making them heavier and harder to place.

Do You Need Engineered Headers?

Engineered-wood components also are used as headers in door and window openings, especially for long spans. They shrink less than wide dimensional lumber, thus reducing cracks and nail pops in drywall at the corners of windows and doors. And insulated headers now are entering the market. But it's difficult to justify the higher cost for small openings unless the added dimensional stability is worth the extra money or unless the header carries an unusual load, such as a column supporting a structural ridge.

"You only want to use it for a larger-span structure, or something with a heavy load on it," says Ed Vanderhoef, a structural engineer in Cheshire, Connecticut. Headers up to 10 ft. long usually can be made less expensively with 2× material, he says.

Scott Gibson, *a contributing editor to* Fine Homebuilding *magazine, lives in East Waterboro, Maine.*

It's difficult to justify the higher cost for small openings unless the added dimensional stability is worth the extra money or unless the header carries an unusual load, such as a column supporting a structural ridge.

Rot-Resistant Framing Material

■ BY SCOTT GIBSON

For 60 years or so, a compound called chromated copper arsenate (CCA) successfully protected structural lumber from the harmful effects of weather, bugs, and microbes. Eased off the consumer market at the end of 2003 due to concerns about its toxicity, CCA was replaced by two other compounds also rich in copper but without the arsenic or the chromium: alkaline copper quaternary (ACQ) and copper azole (CA). Now, nearly four years later, they remain the only residential choices for treated wood rated for ground contact.

Technology in lumber treatment is moving forward, however. Manufacturers are working hard on more-benign organic compounds that not only eliminate the corrosive effects of traditionally treated lumber products but also can be used directly in the ground.

Until that technology is ready for market, three effective structural materials can be used when mold, insects, and decay are a concern. Each has varied applications, levels of protection, and price differences.

Copper-based ACQ and CA are highly recognizable materials that continue to dominate the preserved-wood industry. Borate-infused lumber is not a new technology, but it is a fast-growing, nontoxic alternative to metal-based treatments. And plastics, which typically have been thought of only as decking material, are emerging as another viable structural material.

Each material has significant strengths and weaknesses that are important to understand before being incorporated in any type of building project.

New materials have lots of benefits. Copper- and borate-based preservative treatments have proven effectiveness, but structural plastics are more durable and are made from recycled materials.

Modern copper-based treatments are less toxic but highly corrosive.

ACQ/CA: Alkaline Copper Quaternary/ Copper Azole

ACQ and CA are the two copper-based compounds that have been the industry standard for residential use since chromated copper arsenate was phased out in 2003. Readily available, ACQ- and CA-treated lumber products have established a reliable track record of protection against moisture, mold, and insects, as well as endured the effects of direct ground contact. Don't be misled,

COMPARING OPTIONS

	ACQ/CA	Borate	FRPL
Best Uses	• Sill plates, decks, landscape ties, outdoor structures	• Sill plates, exterior-wall framing, Zone framing in mold-prone locations like sheds, outbuildings, and basements	• Outdoor and marine projects like sheds, playgrounds, decks, docks, and retaining walls
The Good	• Readily available at lumberyards and home centers. • An able successor to CCA lumber. Suitable for both above-ground and in-ground applications.	• Not harmful to humans, pets, or the environment. • Can be disposed of like untreated lumber and even can be burned. • Has no corrosive effect on metal fasteners or hardware. • Can be applied to dimensional as well as engineered lumber.	• Resists decay in even the most severe environments. • Nontoxic and nonleaching. • Requires little maintenance once in place.
The Bad	• Builders required to use pricier hot-dipped galvanized or stainless-steel hardware. • Overall effect on the environment is still unknown, so disposal is a concern.	• Preservative readily leaches out of wood in wet conditions. • Can't be used in ground contact or in areas with high exposure to the elements. • Some manufacturers suggest that cut ends be coated with a borate solution, which slows construction time.	• Load-bearing capacity is less than dimensional lumber of the same size. • It's heavy, expensive, and currently not recognized by the IRC.
Cost	• 2-in. by 6-in. by 8-ft. stud: $8* • 6-in. by 6-in. by 8-ft. post: $22 • 4-ft. by 8-ft. by ½-in. plywood: $38	• FrameGuard: 2-in. by 4-in. by 9-ft. stud: $3 • BluWood: 2-in. by 4-in. by 10-ft. board: $5.50 • BluWood: 4-in. by 8-in. by ¾-in. plywood: $22	• FiberForce 2-in. by 4-in. by 9-ft. stud: $27 • FiberForce 4-ft. by 8-ft. by ½-in. plywood: $113
Sources	• NatureWood® (www.osmose.com) • ProWood® ACQ (www.ufpi.com) • Wolmanized® Outdoor Wood (www.wolmanizedwood.com)	• BluWood® (www.perfectbarrier.com) • FrameGuard® (www.wolmanizedwood.com) • Hi-bor® and Advance Guard® (www.osmose.com)	• American Composite Timbers (www.compositetimbers.com) • FiberForce® (www.plasticboards.com) • Trimax® Structural Lumber (www.trimaxbp.com)

though: Not all copper-based treated wood is intended for in-ground use. Copper-rich preservatives are forced into wood with various retention levels depending on the lumber's intended end use. A label attached to each piece of lumber specifies the wood's proper application.

The transition from CCA to ACQ and CA has not been seamless. Eliminating arsenic and chromium reduced treated lumber's toxicity but significantly increased the wood's corrosive effect on steel fasteners and hardware, as well as on some types of flashing.

Copper corrodes steel—like nails and joist hangers, for example—through a process known as galvanic corrosion. When these two dissimilar metals come in contact and are exposed to moisture, the less-resistant, more-active metal (steel) becomes anodic and literally is consumed by the cathode, or the stronger, more-noble metal (copper).

The increased levels of copper also have made ACQ and CA lumber more expensive, and the absence of chromium allows more copper to leach into the environment. That's a problem for aquatic organisms, and it could lead to government restrictions in the future. As a result, the industry sees ACQ and CA as a step on the way to third-generation treatments.

Borate, a naturally occurring mineral salt, is a great wood preservative in dry environments.

Borate can be infused into many types of lumber.

Borate

Borate is absorbed easily into wood fibers, making it ideal for treating lumber. Borate also has a proven history as an effective agent against mold, wood-destroying microbes, and insects like the Formosan termite.

Among its positive qualities, borate can be infused into the array of lumber used in construction, including I-joists, 2× studs, sheathing, and even LVLs.

Borate has one serious fault, however: a tendency to leach out of wood under frequent water exposure. This quality makes borate unsuitable for outdoor or in-ground use. According to some manufacturers, though, brief job-site storage—up to

six months—won't affect the lumber's protective quality.

Efforts to fix the leaching problem are underway. "The holy grail of wood preservation is a borate that's nonleachable," says Tor Schultz of Mississippi State University's College of Forest Resources. "In the lab, borates can be modified so that they won't leach out of the wood, but as permanence increases, effectiveness drops like a rock."

Until the leaching limitation is overcome, borate-treated lumber will remain a smart choice for locations not directly exposed to the weather or in homes where mold allergies are an issue.

FRPL: Fiberglass-Reinforced Plastic Lumber

Made with high-density polyethylene and fiberglass, this material is available in dimensions from 5/4-in. deck boards to 12×12 posts and even as sheathing designed to replace plywood.

Plastic lumber doesn't absorb water, so it won't crack or rot over time. It does not leach; it resists insect and mold damage; and it resists degradation from oil, fuel, and fungi. Used milk jugs, recycled detergent bottles, and the like make up the plastic component of the lumber, so it's also a green building material. Sound perfect? It might be for high-stress marine and commercial environments where even conventional pressure-treated lumber isn't tough enough.

However, plastic lumber has three major drawbacks: cost, weight, and flexibility. Two or three times as expensive as treated wood, plastic lumber weighs a lot: A 10-ft.-long 2×4 can weigh more than 20 lb. This material is also susceptible to "creep," a tendency to sag under a load over time. As a result, structural plastics are limited to shorter span distances than equally sized dimensional lumber. Residential use of the material is not yet listed by the International Residential Code, so it might be limited to housing projects that carry an engineer's stamp; check with local building officials for approval.
Prices are from 2007.

Scott Gibson, *a contributing editor to* Fine Homebuilding *magazine, lives in East Waterboro, Maine.*

Plastic lumber is extremely durable and environmentally friendly.

Plastic lumber is an excellent choice for docks.

New Compounds and Organic Treatments Are on the Way

The future of wood preservatives most likely belongs to an emerging class of carbon-based compounds.

Chemical Specialties Inc. (CSI), for example, is expected to begin production of an organically treated lumber called Ecolife. According to CSI, the wood will be less corrosive to metal fasteners than wood treated with copper-based preservatives. Ecolife's intended use is outdoors—as decking material, for example—but not for ground contact. However, the company says it expects to have ground-contact preservatives on the market soon.

Developing an effective organic treatment that can be used outside and especially in contact with the ground is no easy task. There are many economical and technical roadblocks in the way. Lengthy government and code reviews are just one of the many inhibiting factors.

Dr. Darrell Nicholas, a professor and researcher at the College of Forest Resources at Mississippi State University, points to a compound called PXTS as a good example of the pitfalls in the approval process. PXTS, an organic sulfur treatment developed by Akzo Nobel, looked promising after years of testing. It showed virtually no toxicity to humans, and it was effective in protecting wood. The company got the compound approved for industrial use, Nicholas says, but efforts to win residential approval were shelved, at least in part because the approval process took so long.

Nicholas says that chemical companies are working vigorously to produce new organic treatments, and he expects several to be available in the next few years.

A company called Timber Treatment Technologies (TTT) is taking another approach with a product called TimberSIL®. Relying on a proprietary combination of sodium silicate and heat, TimberSIL is supposed to be non-corrosive to fasteners, nonleaching, nontoxic, and fire-resistant, and it also is suitable for ground contact.

Distribution, however, is limited, and the company has become entangled in a legal dispute with its original treater. Also, TTT has not publicized independent performance data verifying the claims that have been made about TimberSIL.

Although it has won new-product awards from *Popular Science* and *Metropolitan Home* magazines and the *GreenSpec Directory,* there are still a lot of questions to be answered about TimberSIL.

TimberSIL claims to revolutionize wood preservation, but lacks hard data.

LVLs: A Strong Backbone

■ BY JOHN SPIER

In the bad old days, we used solid lumber for everything. It wasn't long or light enough, but it was versatile. We could "site-engineer" a beam for almost any application. In the past 15 years, engineered-lumber beams have made it possible to build floors flatter, stronger, and more stable than ever before. Selecting the best type of engineered lumber, however, can be a confusing process. I frequently have to slow down my thinking when I am calling the folks at the lumberyard to order LVLs, PSLs, LSLs, or glulams (see the sidebar on p. 88).

For the central bearing beam of a floor system—the backbone of a wood-frame floor—I prefer to use laminated veneer lumber over other types of engineered lumber. With LVLs, I can build up several pieces into a beam large enough to carry almost any load. LVLs combine the versatility of solid lumber with the strength, stability, and uniformity of engineered material.

Laminated veneer lumber lets you assemble big beams without a big crew.

How LVLs Compare to Other Options

Widening the world beyond solid wood and steel, the engineered-beam kingdom offers four additional choices. Engineered lumber is longer, stronger, flatter, heavier, and more expensive than solid wood. Some beams are available pressure-treated for damp or exterior applications. For the few situations where neither solid nor engineered lumber will work, steel is still a great choice.

LAMINATED VENEER LUMBER

The most versatile to work with. Thin layers of wood glued together resemble high-density plywood with parallel laminations. Stronger and stiffer than glulams, LSLs, or solid lumber, LVLs can be built up for long headers or major carrying beams. Width: 1¾ in. Depths: 5½ in. to 20 in., including standard solid-lumber sizes (5¼ in., 7¼ in., 9¼ in., and 11¼ in.) and engineered-joist sizes.

PARALLEL STRAND LUMBER

The strongest and stiffest engineered-lumber beam. Long strips of waste-wood fiber are glued parallel to each other. Widths: 21¹⁄₁₆ in., 3½ in., 5¼ in., and 7 in. Depths: 9¼ in., 9½ in., 11¼ in., 11⅜ in., 14 in., 16 in., and 18 in. Also available as posts and pressure-treated for exterior use.

STEEL I-BEAM

Best strength-to-depth ratio. Available in nearly any size imaginable, steel beams can solve load puzzles where tall wooden beams can't fit. Steel is suitable for exposed interior applications. You also can beef up the load capacity of solid lumber by sandwiching a steel flitch plate between planks.

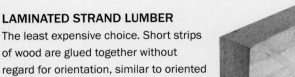

LAMINATED STRAND LUMBER

The least expensive choice. Short strips of wood are glued together without regard for orientation, similar to oriented strand board. LSL is good for door and window headers, including garage-door headers. Also suitable for beams, LSLs aren't as stiff as the others but offer more drilling flexibility. Widths: 3½ in. for beam and header stock. Depths: 4⅜ in. to 16 in., including standard solid-lumber and engineered-joist sizes. LSLs are also available as posts and rim boards.

GLUE LAMINATED TIMBER

The most versatile at the drawing board. Stacked framing lumber is finger-jointed together lengthwise and laminated on top of each other. Glulams can be used in arcs, giving tremendous design flexibility. Laminating a slight camber into a beam can combat deflection. Architectural-grade glulams are suitable for exposed interior applications. Widths: 3⅛ in., 3½ in., 5⅛ in., 5½ in., and 6¾ in. Depths: Up to 50 in., including standard solid- and engineered-lumber sizes.

SOLID LUMBER

The original alternative to full-dimension beams. Solid lumber is still strong enough to handle many load-bearing chores such as headers and floor beams, but it needs more intermediate support than engineered lumber. Solid lumber is also prone to shrinking and cracking.

Dropped Beams for Basements

Carrying beams can be dropped beams (installed under floor joists) or flush beams (installed in plane with floor joists).

Dropped beams make floor framing faster because the joists run over the beam in continuous lengths. Rather than sitting on the mudsills, a dropped beam typically requires a pocket in the foundation wall to support each beam end. Dropped beams are nice because the bays between joists are open, which makes running utilities easier; the exception is where a long sewer line or HVAC trunk line needs to run under the joists and cross the beams. Another drawback to dropped beams is that they reduce the overhead clearance, which may be a code issue if you're planning to turn the basement into living space.

Flush Beam or Dropped Beam?

Two alternatives for placing a beam offer distinct advantages and disadvantages. Typically, dropped beams are located in the basement to support the first floor, and flush beams are concealed in ceilings for second- and third-floor framing.

DROPPED BEAM

PRO
- Continuous joists are faster, stronger, and less prone to floor squeaks.
- Generally less labor intensive.
- Easier to install plumbing and electrical utility runs.

CON
- Less headroom below.
- Requires pockets in foundation or framing.
- Can interfere with large duct runs or plumbing waste lines.
- Often requires blocking above beam.
- Wrapping the beam for finished living space requires extra work.

FLUSH BEAM

PRO
- Fits within floor system to maximize headroom or to maintain an unbroken ceiling plane.
- Rests on foundation; no beam pockets needed.
- Solid bearing surface for point loads.

CON
- Drilling limitations restrict utility runs.
- Labor intensive; requires more cutting and nailing than dropped beam.
- Hangers often protrude below framing plane.
- Prone to floor squeaks.

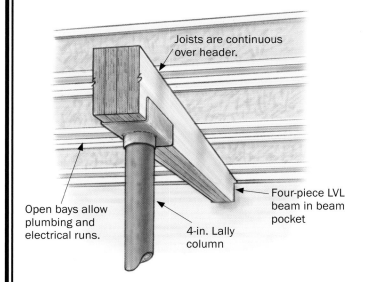

Joists are continuous over header.

Open bays allow plumbing and electrical runs.

4-in. Lally column

Four-piece LVL beam in beam pocket

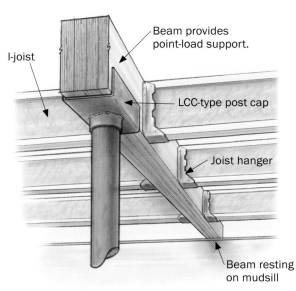

Beam provides point-load support.

I-joist

LCC-type post cap

Joist hanger

Beam resting on mudsill

Beam Ends Need Solid Support

Beam pockets, whether formed or cut into masonry walls, need to be sized and detailed correctly. In a perfect world, accurate plans on-site before the foundation is poured would ensure a perfect pocket. The beam would sit on sill seal and be flush with the mudsill.

In the real world, trimming or shimming is usually necessary. Steel plates and solid masonry materials make great shims, but pressure-treated wood works well, too. Just keep the grain facing up and the blocks longer than about 3 in.

For the inevitable situation when a beam pocket in masonry needs to be enlarged, use an angle grinder outfitted with a diamond blade. Don't use a hammer drill with a chisel bit because you'll likely fracture the foundation at this crucial bearing point.

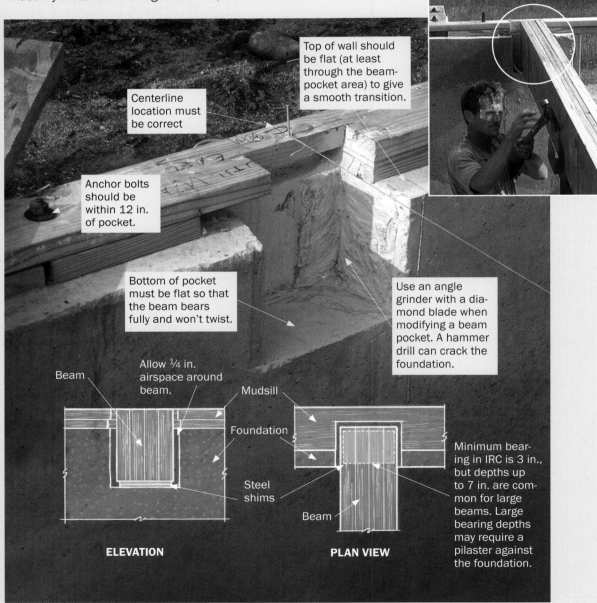

Top of wall should be flat (at least through the beam-pocket area) to give a smooth transition.

Centerline location must be correct

Anchor bolts should be within 12 in. of pocket.

Bottom of pocket must be flat so that the beam bears fully and won't twist.

Use an angle grinder with a diamond blade when modifying a beam pocket. A hammer drill can crack the foundation.

Beam

Allow ¼ in. airspace around beam.

Mudsill

Foundation

Steel shims

Beam

ELEVATION

PLAN VIEW

Minimum bearing in IRC is 3 in., but depths up to 7 in. are common for large beams. Large bearing depths may require a pilaster against the foundation.

Flush Beams Make Flat Ceilings

Flush beams are both labor- and hardware-intensive, but they preserve headroom and the plane of the ceiling below. Flush beams can be built and installed using most of the same techniques used for dropped beams. There are a few extra considerations and opportunities, though.

Flush beams often can be set in place with the hangers already installed, which eliminates a lot of overhead nailing. On big, heavy beams, though, I don't like to preinstall joist hangers. If the beam slips while being lifted into place, an attached joist hanger could hurt somebody.

Support the Middle with Columns and Caps

Flush or dropped, a carrying beam usually is supported at intermediate points by columns. In a basement or crawlspace, this usually means Lally columns on concrete footings, or pads. Because LVL beams are stronger than regular beams, the supports often are farther apart. Greater spans between columns usually mean that larger footings are needed. Sometimes, these point loads require additional reinforcement in or on top of the concrete, too, so it's worth checking with your engineer.

Continued on p. 94.

Column Caps

Some type of cap over a steel Lally column is required by building codes to protect the lumber from being sheared by the metal column. I like the Simpson Strong-Tie LCC-type post cap because it secures the beam tightly in a saddle, distributes the load evenly, and is sized to fit built-up LVL beams. These caps also look good in an unfinished basement.

Made with 12-ga. steel, they're available for beams ranging from triple 2×10s up to seven-layer LVLs and for 3½-in. or 4-in. Lally columns. The one pictured here, sized for three LVLs and a 4-in. column, weighs 5½ lb. and costs about $46*.

Building Up the Beam

2 Build up the beam one layer at a time. Slide additional LVLs along the top of the first one, then drop them into the saddle of the post cap.

1 **Work from the top down.** Set the post cap first. Tack the cap to the first layer of LVL, then slip the column into the cap. Slide the column into place on the footing below. Adjust the column for plumb from the bottom.

Spike first, then nail it off with a gun. Nail guns put a lot of nails in a beam in a hurry, but they don't squeeze the layers together as well as hand-nailing does.

Fill every hole with a nail. All Simpson Strong-Tie LCC-type post caps require eight 16d nails (four per side). Use full-length 3¼-in. 16d nails rather than 1½-in. joist-hanger nails.

To cut the columns, use a large pipe cutter or a reciprocating saw to cut through the steel, and then break the concrete filling by whacking the cut with a hammer.

If the beam goes through or over an intersecting wall, it generally can be supported in the wall with multiple jack studs or an engineered PSL post.

Build Beams One Piece at a Time

A one-piece beam simply can be cut to length and dropped into place. If it is delivered on a crane truck, you often can persuade the crane operator to set it in place for you, but you need to respect his time. Work quickly to cut the beam to length, and have the beam pockets ready to go. A chainsaw can make cutting big beams go faster.

Because I work with a small crew, I usually choose to build up beams from more manageable layers of LVL, which are slid into place and spiked together. There are no crowns in LVLs (as with solid lumber), but I keep the lettering right side up to make my customers happy.

It's important to hand-nail the layers together before letting loose with a nail gun. Nail guns won't squeeze together the layers, but 16d sinkers and a 28-oz. framing hammer will. Even if the manufacturer's fastening schedule includes lag screws or bolts, hand-nailing first is a good idea. One more thing to keep in mind: Keep the pieces relatively straight as you nail them, or you can nail a curve into the beam.

Level, Square, and Secure the Beam

After the beam is assembled and in place, it needs to be checked for level and secured into its final position. Make sure the top of the beam planes smoothly into the mudsills, then use a transit, laser, or carpenter's level to check that the middle of the beam is flat and level.

Most building codes require built-up carrying beams supported by columns to be contained in column caps. I usually use an LCC-type cap manufactured by Simpson Strong-Tie Co. Inc. (www.strongtie.com; 800-999-5099). The saddle that is created by the cap secures the beam, ensures that the load is distributed evenly, and meets my local code requirements.

When using post caps like these, you need to incorporate them into the beam-building process because you can't slip them under the beam after it is assembled, or at least not without jacking up the beam. To determine the column height, measure down from a string pulled taut over the mudsills (see the sidebar on p. 90). Take the height from the footing to the string, and then subtract the beam depth and the thickness of the plates on each end of the column. Cut and label the columns. I add about ⅛ in. in the middle of a 32-ft. span to allow for string sag and column-base compression.

Height issues need to be addressed now, before loading the beam with floor joists and plywood. If intermediate supports are too high, they can be trimmed; if they are too low, they can be replaced or shimmed with correctly sized steel plates that will handle the compressive loads. Many framers set solid-lumber beams high to allow shrinkage and compression. Don't do this with LVLs, though, or you'll build a hump into the floor. LVLs won't shrink or compress the way that solid lumber will.

Check that the beam is placed squarely, especially if it defines a stair opening; then secure the ends of the beam to the mudsills with short pieces of metal strapping. Check for straightness with a careful eye or by pull-ing a string along one corner. Temporary braces can hold the beam straight while you secure it with the floor joists.

Prices are from 2005.

This chapter was adapted from **John Spier's** book, Building with Engineered Lumber *(The Taunton Press, Inc.).*

Use Studs to Make a Wall Pocket

When the beam ends short of a foundation wall, such as in a framed wall defining a stair opening, a post can be built up within the wall using studs. Nail a full-height stud into the side of the beam for stability.

Factory-Framed Floors

■ BY FERNANDO PAGÉS RUIZ

I pride myself on being the most efficient builder possible. For nearly 15 years, I've built homes using factory-framed components to improve production efficiency in my company. With factory-framed walls and roof trusses, I thought the industry had reached its peak in framing technology. So when my wall-panel manufacturer suggested testing a new factory-framed floor system, I expected only an incremental improvement in production time. I didn't know that I would come to regard factory-framed floors as perhaps the most important framing enhancement since factory-made roof trusses.

My first experience with factory-framed floors was on the PATH Concept Home in Omaha, Nebraska, a demonstration house that my company built. This house boasts the latest technology and building methods in residential construction. Weyerhaeuser's iLevel NextPhase® program (www.ilevel .com), which includes this floor system, was an important element in that project. It showed how off-site, component-built floor panels could improve quality and safety on the job site while greatly reducing construction time.

During the building process, I expected that a solid portion of a day would be devoted to laying the first floor. However,

Factory Plans Are More Refined

The plans, created by iLevel software, are used by factory crews to build floor panels that include HVAC, plumbing, and framing details. The factory set of plans also includes the proper stacking order for delivery and installation information for the on-site carpentry crew.

Rim boards are labeled with two numbers. The first indicates the panel the board belongs to (P1), and the second indicates its positioning in the layout (R16).

HVAC and plumbing openings are labeled on the plan, but arrive on site already cut in the floor joists.

Beam locations are marked prominently on the layout plan.

Red triangles indicate the direction each joist is to be fed into the saw and in what direction it should be assembled in the panel.

Crane-lift holes are marked at precise points on each panel to ensure that the panels fly level.

Floor joists have the same labeling convention as rim boards. The first number indicates the panel that the joist belongs to (P1), and the second lets the assembly crew know where to position the joist (J3).

Computer software creates the truck's stacking order. The stack of panels should arrive on site with the first panel on top and the last panel on the bottom.

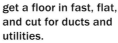

Crane-hoisted panels get a floor in fast, flat, and cut for ducts and utilities.

Floor panels are built in a factory setting. Trained crews assemble each panel on flat, square framing tables to maximize production speed and quality control.

the carpenters arrived on site at 8 a.m., and by 9:15 the first floor was complete. The first floor would have been done sooner, but the media covering the PATH project delayed the process. I was told that without any interruptions, the floor would have been done in about 30 minutes. In any event, I was impressed.

Panelized Framing Components Are Not New, but Floor Panels Are

Because they're so new, these floor panels can be hard to locate. In fact, iLevel is the only company I could find that produces them. iLevel suggests going to their website to find the nearest mill or lumberyard that has been set up to build floor panels with the system.

Floor panels come in 8-ft. widths and in up to 60-ft. lengths. Each floor panel typically is constructed with engineered I-joists or open-web trusses, LSL rim boards, and oriented-strand-board sheathing. In addition to being more stable, these engineered-lumber components are more uniform and are more consistent than traditional dimensional lumber.

The floor systems I use are designed with proprietary iLevel software. The programs control machinery that measures and marks each component with cut locations, sheathing layouts, and nailing patterns. Floor joists also are marked for HVAC and plumbing chases. To minimize waste, the software keeps track of cutoff material and configures usable pieces back into other panel plans.

When the components are marked, computer-operated saws and trained crews cut and assemble each panel on perfectly flat, square framing tables. The assembly takes place in a factory, which has a major benefit. The floor is protected from the weather, so building schedules can stay on track despite days of driving rain or lingering snow.

Coordinate with Subcontractors Early in the Construction Cycle

To take full advantage of the floor-panel system, start by refining building plans on paper with your subcontractors. Make sure the blueprints accurately represent the house you want to build. Ask the plumber and the HVAC contractors to draw their piping and ducting scheme on the plans, with pipe and duct sizes noted.

A team of factory engineers further refines the plans and will alert you of any structural discrepancies. Sometimes the duct-layout plan has a plenum too large for the floor system you have chosen. Have the subcontractor find another scheme within the limitations of the floor system, such as running two ducts instead of one, or deal with ducts the conventional way by using a dropped ceiling.

Once the factory engineers have gone over the plans, they provide you with the dimensions needed to pour the foundation.

Foundations Have to Be Absolutely Square and Level

To use any factory framing system effectively, including panelized walls and roof trusses, foundations have to be near perfect. A foundation 1 in. or more out of square over 60 ft. is a problem. Of course, you should expect a high level of accuracy even if you frame conventionally. But trying to place a perfectly square floor panel on a crooked foundation is like trying to pound a square peg into a round hole. For this reason, a factory rep usually measures the foundation before the floor is built.

If the foundation discrepancies exceed factory tolerances, the rep might call you with the bad news: "You'll have to stick-frame this one." Josh Weekly, a framing supervisor for Millard Lumber in Omaha, has made this call more than once.

A visit from the factory rep isn't a requirement, though. "With experience, you can knock a few days off the construction calendar by ordering the floors based directly off your plans—at your own risk," says Weekly. However, when my company is working with a new floor plan, I always have a factory rep take foundation measurements. Once the rep has the dimensions, the rep goes back to the factory and panel assembly begins. This process takes a bit longer, but I don't worry about the floor not fitting the foundation. The panels usually are ready about a week after the rep's visit.

While the Panels Are Being Built, Prep the Foundation

"Laying the sill plate is the most important step in the process," says Weekly, who often trains new framers on using floor panels. It's not much different than plating a conventional framing job. However, a seamless panel installation hinges on the accuracy of the sill-plate layout (see the sidebar below).

Weekly rarely looks at blueprints during this process. "Only for reference," he says, adding that "if you follow the factory version of the layout plan exactly, the floors will fit perfectly, with bearing points

---

A Simple Way to Lay Sill Plates Precisely

The mudsills should be straight and square, even if the foundation walls aren't. Considering the precision required when installing factory-framed floors, you shouldn't assume that the foundation has been poured perfectly. Our mudsill layout doesn't involve setting the sill a uniform distance from the inside or outside of the foundation. Instead, we establish corner points based on the dimensions labeled on the factory plans. Factory dimensions are to the exterior face of the sill plates, about ½ in. from the edge of the foundation to provide space for the wall sheathing. Here's how we lay out a standard 2×6 mudsill when working with factory-framed floors.

Step 1: Check the plan measurements. Wall and diagonal dimensions should be prominently labeled on the plans. They'll be referenced to establish all the corner points.

Step 2: Start with a long wall. Establish points A and B by measuring ½ in. in from the outside of the foundation corners. Snap a line between points A and B.

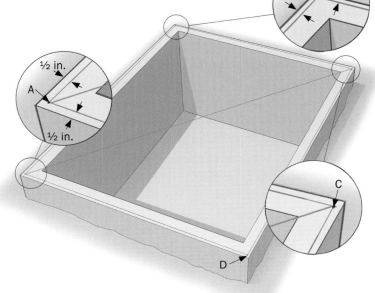

Step 3: Mark point C by intersecting measurements BC and AC found on the layout plan to form a right triangle. Snap a line from point B to point C.

Step 4: Mark point D at the intersection of measurements AD and CD, also found on the layout plan. Snap a line from point A to point D and from point C to point D.

Step 5: Double-check all the measurements and both diagonal distances. The diagonal measurements should be within ⅛ in. of each other. If for some reason the difference exceeds ⅛ in., adjust line CD until the diagonals match, and resnap the short-wall lines.

centered on walls and beams, plumbing and duct openings lined up like gun barrels, and stairway openings, cantilevers, and insets balanced perfectly in relation to the foundation and the wall framing to come."

In addition to the sill-plate installation, basement bearing walls need to be built. "We stick-build most basement bearing walls on site to take up any slack in the slab level," says Weekly. Carpenters can cut the wall ¼ in. to ½ in. low to make up for waviness in the basement floor. Later, they shim between the floor panels and the top plates for a level installation.

To check for level, carpenters pull a stringline from sill plate to sill plate over the bearing-wall and beam locations to make sure they do not stand proud. Panels are square and level. If you have a wavy foundation or uneven interior bearing points, the floors will not bear fully.

A Crane Moves the Panels

The morning the floor panels arrive, the mood on the job site resembles the mood on a concrete pour. Anticipation builds by the minute, with all hands ready to hit the deck for a burst of furious activity. A crane arrives, parks in the street a few minutes early, and drops its outriggers. The boom is extended, ready to unload panels and hoist them into place. My lumberyard, Millard Lumber, supplies a crane with every delivery, given that there's no other way to unload and move floor panels. They're too big for an all-terrain forklift and too heavy for even the burliest of crews to maneuver. At a minimum, you need a 20-ton crane with a 60-ft. boom, long enough to pick up panels and place them on a structure two or three stories high. Crane fees can vary, but they cost about $150* an hour here in Omaha, where I build.

A semi with a flatbed trailer pulls up with a load of panels stacked in the proper order of construction: first panel on top, last panel on the bottom. However, stacking and loading requirements occasionally force a change in the arrangement, so the first order of business involves checking the panel numbering.

Lay a bead of construction adhesive along the panel's end joist.

Fly the next panel into place, shiplapping the sheathing over the rim joist and the end floor joist.

Every panel is keyed to the plans with a number, so it's easy to see if no. 1 rests on top or three panels down. If panels are out of order, the crane operator can sort them. The stack of panels, organized by the same software that engineers the floors, usually works seamlessly with the construction sequence.

To move the panels, the crane operator uses bands of reinforced fabric strapped to balance points on the joists. Again, software locates the "lift holes" for each panel, and a 4-in. hole saw drills through the sheathing directly over the proper point. On site, a crew member threads a strap through each hole, loops it around the floor joist, then slips the strap over the crane's hook. The panel stays level as it's maneuvered into place.

Placement of First Panel Is Critical

Like setting the bottom course of a masonry wall, the first panel determines square and true for all the rest. "It's mainly a question of lining up the three sides of the panel's rim joists precisely with the sill plate," says Weekly. After lowering the panel as carefully as possible, adjustments are made by tapping the panel with a sledgehammer until each corner lines up within 1/16 in. of the factory specification. After measurements confirm accurate placement, a few toenails through the rim board and into the sill hold the panel in place. As subsequent panels are lifted, carpenters lay a thick bead of construction adhesive along the edge of the previous panel's end joist. Sheathing on every panel except the first is set back 1 9/16 in. from the edge of the end joist to shiplap the panels together. A few gentle taps with a sledgehammer butt the panel tight to the previous one, and a sharp eye makes sure the panel is parallel with the sill plates.

Some panels can take a bit of persuasion to come together. Carpenters draw these panels together with a wall puller, a tool that has two pick ends that dig into the floor sheathing with a lever that, when pulled, ratchets the panels tight. Originally designed to pull a framed wall into alignment, this cool but nonessential tool in the conventional framer's kit becomes an indispensable piece of equipment for panelized-floor installation.

As the next panel flies off the truck, the carpenters double-check measurements from the edge of the rim joist to the end sill plate. They do this after every panel placement before toenailing the rim joist to the sill.

TIP

Hard hats are a must when working around a crane. Don't assume otherwise just because this builder forgot his.

3 Make sure the panel is square to the sill or top plates, and pull it tight to the previous panel.

4 Nail the panel to the floor joist and into the sill or top plates.

Second-floor safety. Installing factory-framed floors on second and third stories is easier and much safer than stick-framing. Carpenters work off a solid floor rather than having to walk the tops of narrow walls. This prevents on-site injuries and increases building speed.

The last panel is usually the smallest. It has a "flying edge" with no end rim joist attached. This allows carpenters to trim the panel to fit the foundation, fixing any small discrepancies that might show up in the final fit between the floor and the foundation walls. To set up the last panel, carpenters measure from each corner of the previous panel to the edge of the foundation's sill plate. They transfer these measurements to the last panel. As the crane holds the panel at waist height, carpenters cut the sheathing and the rim joist to the exact measurement. "It's easier to trim the panel suspended at table height," says Weekly. However, the panel can be adjusted in place if needed.

After the last panel is in place, it's time to plug the lift holes left in the sheathing. First, dried-glue residue from manufacturing is removed. Then, with 4-in. round cutouts

Avoiding Problems

The majority of problems encountered when working with factory-framed floor panels are due to human error. Mismeasuring, entering the wrong data, and installing the panels the wrong way all can create havoc on the job site. If you check the foundation for square but not level, don't be surprised when the panels don't lie flat on the mudsill. If the carpenters install a seemingly identical panel in the wrong order, they could discover later that a key bearing point was missed entirely. Here are a few tips to avoid the most-common problems:

- Compare blueprint details with panel CADs. Pay particular attention to beam pockets, stairway locations, and other critical measurements. Report inconsistencies to the factory before ordering the panels.
- Follow the engineered drawing dimensions to $1\frac{1}{16}$ in. when installing sill plates and laying the first panel.
- Install panels in the sequence indicated by the factory. Panels look alike, but do

not install like roof trusses, where order doesn't matter.

- Check the labels on every panel to make sure you orient them properly during installation (right and left sides). One panel out of sync can compromise the rest of the layout (especially with duct and plumbing runs). This problem is hard to fix once the crane has left the job site.
- Keep top plates and sills level. Whether load-bearing or not, interior walls should have their top plates level with the mudsill (on the first floor) or the top plates of exterior walls (second floor). Otherwise, the floors will teeter-totter over the high spots.

supplied by the lumberyard, the plugs are glued and screwed in place. Once the floor is laid, one worker can toenail the joists to the sill plates while the rest of the crew starts snapping lines for wall layout.

Laying Second-Story Floors Mirrors the Installation of the First

The safety advantage of using factory-framed floor panels really comes to light as you move higher up the structure. You don't have to hump joists and sheathing up ladders, and you can do your work from a solid platform rather than taking the risk of having a floor joist roll underfoot.

Laying the panels on walls rather than on a concrete foundation gives you greater flexibility to adjust for variances between stories. Just as with the beams and bearing walls in the basement, it's critical that the top plates stay in plane from wall to wall so that the floors lie perfectly flat. Because the second-floor dimensions are typically the same as the first, simply plumb the exterior walls and nail the panels through the rim joist to the top plates.

Panels Are More Valuable Than Cost Savings

For an exact price comparison between factory- and site-framed floors, you would have to frame the same floor twice, once in the field and then again in the factory. Even if you could afford to take on this comparison project, you would have to decide when to do it: during a perfect summer day, in the rain, in bitter cold, with or without wind? Each climatic condition would influence the result—out in the field. One of the difficult-to-quantify but obvious advantages of the factory floor comes with climate and quality

control. Have you ever tried building one cabinet in a well-organized shop and another in a driveway full of ruts and loose gravel?

Although computer-operated machines mark and cut all the joists and beams to exact dimensions, the assembly work for factory floor panels is done a lot as it is on the job site, with carpenters rolling joists and nailing sheathing. The difference is that these carpenters are working at a comfortable height, on level, square framing tables instead of balancing on top of second-story walls. Also, the factory floor is always dry and warm. On average, it takes factory hands about 30 percent less time to build a floor in these conditions than it would take the same carpenters to roll joists on the job site.

Once the floor panels arrive at the job site, the savings in time is dramatic. A true cost comparison depends on what you pay framers and how you quantify material waste and handling. At the factory, scrap lumber is used regularly; on the job site, a lot of it goes into the trash. Your time at the job site includes costs such as interest, insurance, taxes, and worker's-compensation fees. The less time you spend on the job (in man-hours as well as calendar days), the lower these overhead costs are. It just depends on where you want your dollars to go: the carpenters or the supplier. I have enough work to keep them both busy. But I would rather have my carpenters moving on to another project than camped out rolling joists.

Prices are from 2007.

Fernando Pagés Ruiz, *a frequent contributor to* Fine Homebuilding *magazine, operates Brighton Construction Company in Lincoln, Nebraska.*

Mudsills

■ BY JIM ANDERSON

Framing a traditional house begins at the mudsill; it's the first piece of lumber that is attached to the foundation. If you build on a foundation that's out of square or level (and they're common), correcting the problem at the mudsill stage will make for a lot less trouble later. The first step in installing mudsills is determining if the foundation is square.

Checking for a Square Foundation

While my helper sweeps off the foundation and checks to make sure the anchor bolts are plumb, I look over the plans for the foundation's largest rectangle. It will provide an ongoing reference for establishing bump-outs (areas outside the large rectangle) and recesses (areas within the large rectangle) that are square to the house and to each other. The illustration on pp. 106–107 shows how to square the large rectangle or to create a large 3-4-5 triangle if a rectangle can't be found.

After squaring and marking its corners, two of us snap chalklines for the large rectangle and any bump-outs or recesses, while a third person spreads pressure-treated 2×4s (or 2×6s if requested) around the foundation to serve as mudsills. Working as a team with a systematic approach is really important on these projects. We begin at the front corner and run the material along the chalkline from end to end, and then do the same in the rear. We fill in the sidewalls last.

Even if the foundation is not square or level, getting the mudsills right will get you back on track.

Before you can install the mudsills, you have to make sure that the foundation is square. And occasionally, foundations are a little out of square. To get the framing off to a good start, you have to snap a series of square layout lines for the mudsills. Locating and snapping lines on the foundation's largest rectangle provides a square reference for the remaining areas lying inside or outside the large rectangle. If a large rectangle can't be found, a large right triangle (see "Finding Square with Triangles" on p. 108) will work.

1. FIND OUT IF THE FOUNDATION IS SQUARE

A. Locate the foundation's largest rectangle and snap a reference line on one of the long walls $3\frac{1}{2}$ in. from the outer edge of the foundation wall. (We're using 2×4 mudsills in this example; for 2×6s, snap the line $5\frac{1}{2}$ in. from the outer edge.)

B. On the wall opposite the reference line, snap a parallel line $3\frac{1}{2}$ in. from the foundation's outer edge. Make sure this line is parallel to the reference line by measuring between them at each end. If they're not and the difference is less than $\frac{1}{2}$ in., simply move the end of the parallel line that measured short until the measurements are equal.

C. To find the rectangle's corners, mark points a, b, c, and d $3\frac{1}{2}$ in. from the edge of the foundation. Make sure that line ab is equal in length to cd.

D. To check for square, measure from points a to d and from points b to c. There's usually some adjustment required, but if you're lucky and the measurements are the same, the foundation is square. Snap the large rectangle's remaining two chalklines.

2. IF THE FOUNDATION ISN'T SQUARE

A. We know that lines ab and cd are parallel, so the problem is in the other two walls (ac and bd). Leaving line ab in place, square the layout by moving points c and d an equal distance toward the corner with the shorter diagonal measurement.

B. Check the diagonals again and repeat the process until the two measurements are equal (within $\frac{1}{8}$ in. is OK).

C. If there's more than an inch difference in the diagonal measurements, adjust the entire layout by splitting the difference among all four corners of the rectangle (so that the mudsill won't overhang the foundation too much). If, after this is done, the mudsill overhangs the foundation by more than $\frac{5}{8}$ in. in any one place, you've got a bigger problem and might want to call your foundation contractor.

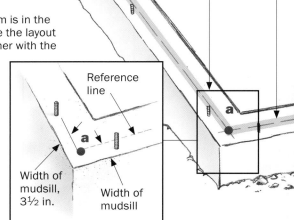

Largest rectangle

Reference line

Width of mudsill, $3\frac{1}{2}$ in.

Width of mudsill

3. LAYING OUT THE RECESSES AND BUMP-OUTS AFTER SQUARING THE LARGE RECTANGLE

Recess

Mark the lengths of each side by measuring from the main rectangle. Snap connecting chalklines.

Bump-out

A. Find the parallel line for the bump-out's outer wall by measuring from the large rectangle (as in step 1B).

B. Measuring from the closest corner of the large rectangle, mark points g and h.

C. On the outer wall, mark points e and f 3½ in. from the edge of the foundation. With each corner now marked, check for square by measuring between e and h, and between f and g. Follow the remaining steps in 1D.

D. If the bump-out doesn't have an outer parallel wall (maybe it's octagonal or circular), you can use the 3-4-5 method to find one of the two perpendicular walls, and use it as a reference line to find the other.

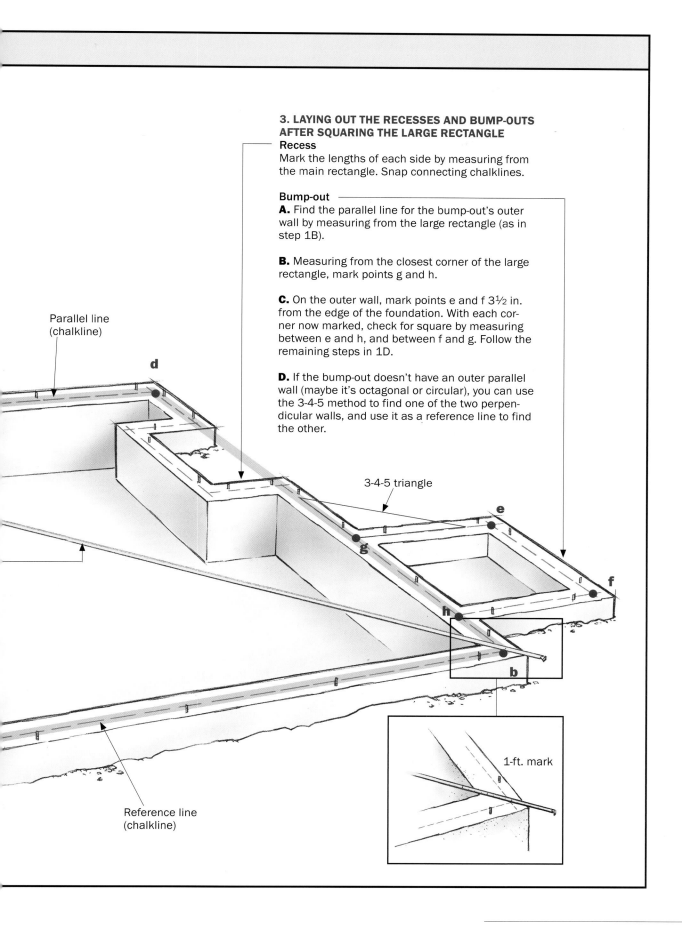

Parallel line (chalkline)

3-4-5 triangle

Reference line (chalkline)

1-ft. mark

Drill holes in the mudsill as straight as possible. An angled hole will pull the plate off the chalkline. Use a ⅝-in. bit for a ½-in. anchor bolt; place a piece of scrap lumber beneath the plate to protect the bit, or cantilever the mudsill beyond the foundation.

Stand the sill plate on edge and trace the outline of the bolt onto the plate.

Finding Square with Triangles

According to the Pythagorean theorem ($a^2 + b^2 = c^2$), any triangle with sides that measure 3-4-5 (or any multiple of these) will always have a right angle opposite the hypotenuse (side that measures 5). If a = 3, b = 4, and c = 5, and $3^2 + 4^2 = 5^2$, then 9 + 16 = 25.

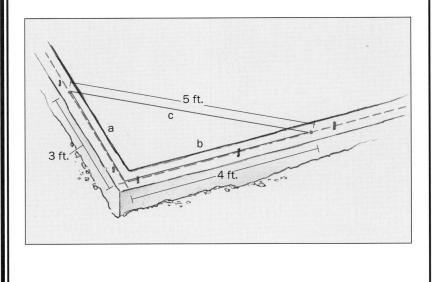

As we work our way around the foundation, we mark the bolt locations by standing the plates on edge and outlining the bolts on the sill plate (see the photo above left).

When we have to join two mudsills, we cut the first plate within 12 in. of a bolt and add an expansion bolt for the adjacent plate. Local code requires an anchor bolt within 12 in. of the end of mudsills or of any joints.

Marking the bolt centers on the mudsills for drilling is next (see the photo above right). It's as simple as laying the mudsill alongside the chalkline on top of the foundation, measuring from the chalkline to the center of the bolt, and transferring the measurement to the top of the mudsills (see the illustration on the facing page). At this point, we add insulation (or sill seal if requested) between the foundation and the mudsill.

A Builder's Level Finds the High Spots

Commonly though wrongly called a transit, a builder's level rotates only horizontally; a transit rotates both horizontally and vertically. Looking through a builder's level is like looking through a rifle scope, cross hairs and all. Properly set up, the horizontal cross hair represents a level plane, and the magnification is great enough to read a tape measure held 100 ft. away or more. A builder's level is leveled with either three or four thumbscrews and integral bubble vials. Comparing measurements taken in different spots tells you their relative elevations. But this comparison can be counterintuitive. The highest spot, being closest to the level's plane, will have the shortest measurement.

For years, I have used a builder's level to install mudsills, and although I have tried laser levels, I haven't been happy with the results. The Sokkia® E=32 level that I now own cost around $400 in 1999, and it has given me great service.

SET-UP TIPS FOR BEST RESULTS

1. Position the level so that you clearly see each of the foundation's corners within a relatively narrow field of view (90 degrees or less). This helps to eliminate errors associated with swinging the level in wide arcs. Place the level as low to the foundation as possible. Extending the tape or measuring rod high in the air introduces error.

2. With a helper holding a tape, shoot the outside corners a, b, c, and d write their elevations on each corner. The shortest measurement is the high corner (b).

3. Subtract the shortest measurement from each of the other corners, and write the difference (the amount to be shimmed) at each corner.

4. Shim the corners until they measure within $\frac{1}{16}$ in. of the high corner.

5. Run stringlines from one corner to the next. For the areas between the corners, see the sidebar on p. 109.

Remove high spots with an air chisel. If a high spot is really bad and it's a short length of wall, an air chisel makes quick work of a labor-intensive job.

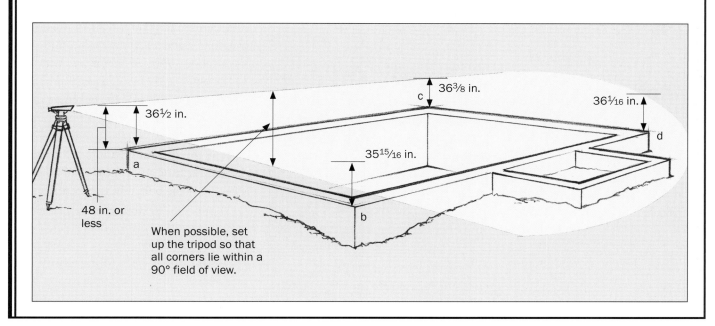

36½ in.

36⅜ in.

36¹⁄₁₆ in.

35¹⁵⁄₁₆ in.

48 in. or less

When possible, set up the tripod so that all corners lie within a 90° field of view.

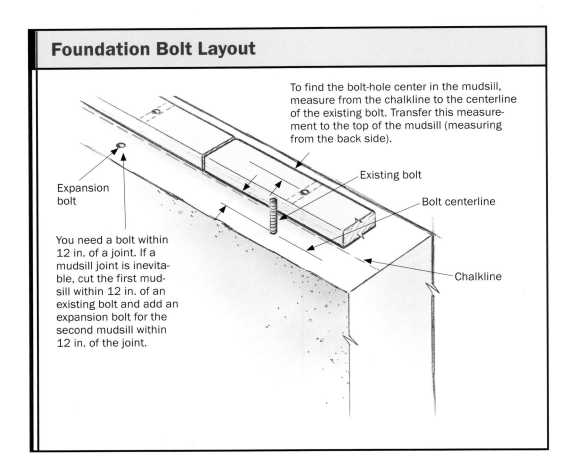

Foundation Bolt Layout

To find the bolt-hole center in the mudsill, measure from the chalkline to the centerline of the existing bolt. Transfer this measurement to the top of the mudsill (measuring from the back side).

Expansion bolt

You need a bolt within 12 in. of a joint. If a mudsill joint is inevitable, cut the first mudsill within 12 in. of an existing bolt and add an expansion bolt for the second mudsill within 12 in. of the joint.

Existing bolt

Bolt centerline

Chalkline

Shims Level the Mudsill; Bolts Hold It Down

After drilling the bolt holes, two of us place the mudsills over the anchor bolts and another follows behind, adding nuts and washers, tightening them only enough to check for obvious high or low spots. Then we add the necessary expansion bolts at the mudsill joints and nail a second 2×4 on the mudsill. This adds an extra 1½-in. ceiling height in the basement.

Next, we use a builder's level (see the sidebar on p. 110) to measure the height of the corners and to look for any high spots. After comparing the measurements, we shim the corners to within 1/16 in. of the highest point on the foundation. Then we run a string from corner to corner and level the mudsills between.

When shims are necessary, the local building code requires steel shims at joist, beam, and point loads, so I mark these locations on the mudsill. After inserting the shims between the foundation and mudsill, we tighten the nuts on the anchor bolts and check the height one last time, shooting for plus or minus 1/16 in.

Jim Anderson is a framing contractor living in Littleton, Colorado.

Shim with Steel

Once you've established a level stringline, use steel to shim between the mudsill and foundation beneath all joist, beam, and point loads.

Sources

Metalwest
1229 S. Fulton Ave.
Brighton, CO 80601
800-336-3365
www.metalwest.com
Steel shims

Sokkia
16900 W. 118th Terr.
Olathe, KS 66051
800-476-5542
www.sokkia.com
Builder's levels

For a rough count, stack the shims up to the stringline in each location. Steel shims are available in 50-lb. boxes.

The second mudsill plate raises the basement ceiling by 1½ in.

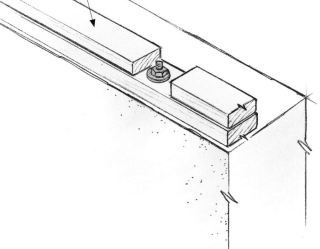

The Well-Framed Floor

■ BY JIM ANDERSON

Walking across a newly framed floor for the first time is a milestone in any framing project. Finally, there's something to stand on that doesn't squish beneath your boots. It's flat and strong, and because there's a floor to stand on, the rest of the project will move ahead much more quickly. But whether you're using common lumber or I-joists (see the sidebar on the facing page), it takes a well-coordinated effort to get any floor to the point where you can walk on it.

Before you start driving nails, it's important to collect as much information as possible about the locations of the joists, posts, beams, point loads, cantilevers, plumbing vents, drains, and HVAC ducts on the floor-framing plan. Whether those details come from the architect, you, or somewhere else, the floor-framing plan needs to reflect the house as it's going to be built.

Having all this information in one place allows you to overlay—in pencil—the big immovable parts of the house on top of each other. This step will catch most if not all the big mistakes that can be made early on. It's a lot easier to erase than it is to remove and replace.

Transfer the Details from the Plans to the Mudsills

First, I check the joist spacing on the floor-framing plan, usually 16 in. or 19.2 in. o.c. and transfer that to the mudsills. Measuring from the end of the house (usually beginning with the longest uninterrupted run), I mark the edge of the first joist 15⅛ in. from the end for 1¾-in. I-joists (16 in. minus half the joist thickness). This places the center of the first joist at 16 in.

Then I mark 16 in. o.c. (or whatever the proper spacing is) from the first mark to the other end of the house. I do this on the front and back walls, then I check the layout marks on both ends to make sure that they are the same. If they are within ¼ in., I leave it; if not, I double-check the layout and make adjustments. I also mark the location of stairs, load-bearing members, and cantilevers on the mudsill.

Why I Prefer I-Joists over Solid Wood

I remember the first time I saw I-joists, those long, floppy things. They seemed so flimsy and light that I thought they would have trouble holding up the sheathing, not to mention the walls that would go on top of them.

They have more than proven me wrong, however. The main advantages are that I-joists are dimensionally stable and very straight. The web (the wide middle section) of an I-joist is cut from oriented strand board, thin strands of wood oriented in the same direction and glued together. Because glue surrounds all those strands of wood, you can expect less shrinking and swelling and very consistent joist sizes (usually within 1/16 in.).

You also can cut much larger holes into I-joists than into solid lumber; holes up to 6 in. are allowed in the center of the span of a 9½-in. I-joist. Elsewhere along the web, 1½-in. holes are provided in perforated knockouts. Holes in solid lumber can be no more than one-third the total width.

I-joists must be handled carefully; upright is best, or supported in a couple of places if carried flat. They're light, come in lengths up to 60 ft., and can span long distances as part of an engineered floor system. Best of all, they cost about the same as lumber; in the longer lengths, they actually cost less.

Whether it's the floor of a big house or a small addition, an accurate layout and efficient techniques promote smooth installation.

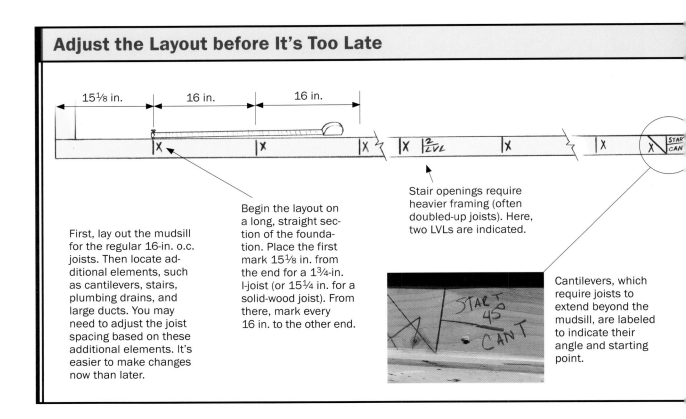

Adjust the Layout before It's Too Late

15⅛ in.   16 in.   16 in.

2 LVL

STAR CAN

STARTO 45 CANT

First, lay out the mudsill for the regular 16-in. o.c. joists. Then locate additional elements, such as cantilevers, stairs, plumbing drains, and large ducts. You may need to adjust the joist spacing based on these additional elements. It's easier to make changes now than later.

Begin the layout on a long, straight section of the foundation. Place the first mark 15⅛ in. from the end for a 1¾-in. I-joist (or 15¼ in. for a solid-wood joist). From there, mark every 16 in. to the other end.

Stair openings require heavier framing (often doubled-up joists). Here, two LVLs are indicated.

Cantilevers, which require joists to extend beyond the mudsill, are labeled to indicate their angle and starting point.

Leave Room for Pipes and Ductwork

If the layout mark for the last joist is within a foot of the endwall, I move it to allow room for plumbing, electrical, or HVAC in what is often an important joist bay. I usually just measure and mark 16 in. from the edge of the mudsill back toward the center of the house.

I also make sure that none of the plumbing fixtures or flue chases lands on a joist. This is another opportunity to double-check myself. It's a lot easier to move the joist now than it is to move it later or repair damage from a determined plumber with a chainsaw. I usually allow a minimum of 12 in. between joists for furnace flues, which provides 2 in. of clearance on each side for an 8-in. furnace flue. Even though 1 in. on each side meets the building code here in Denver, I figure that where heat and wood are concerned, more room is better.

Again, I create this space either by moving the joist off the 16-in. o.c. layout or,

when that isn't practical, by cutting the joist just short of the flue and supporting it with a header tied into the joists on each side of the one that's cut.

Plumbing drains and supply lines are zero-clearance items, so I can have wood right next to them. I locate the fixtures on the plan, and if a joist is on or near the centerline of the drain, I move the joist 1 in. or 2 in. in one direction or the other. If I have two fixtures close together and moving a joist away from one drain places it beneath another, I open the spacing a little more (and double the joist) so that both drains lie within a slightly oversize bay.

Prepare Material According to Where It's Needed

Wood I-joists come from the yard in a large bundle; the rim material and any LVLs usually are strapped to the top. With a helper,

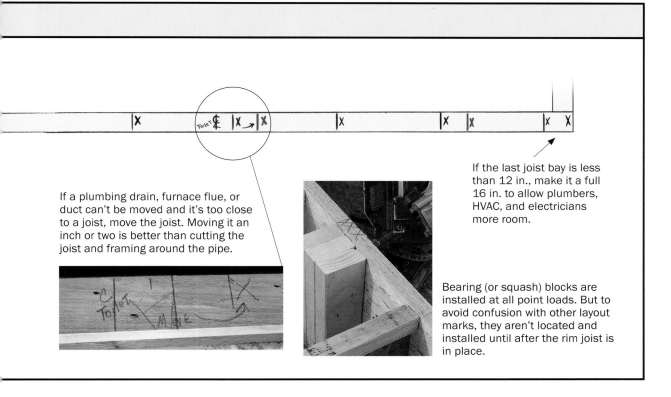

If a plumbing drain, furnace flue, or duct can't be moved and it's too close to a joist, move the joist. Moving it an inch or two is better than cutting the joist and framing around the pipe.

If the last joist bay is less than 12 in., make it a full 16 in. to allow plumbers, HVAC, and electricians more room.

Bearing (or squash) blocks are installed at all point loads. But to avoid confusion with other layout marks, they aren't located and installed until after the rim joist is in place.

I move the LVLs to sawhorses for cutting to length and to install joist hangers.

We move the rim joists to the top of the sheathing or to the ground, and place stickers beneath so that we can lift them easily later. Then we square one end of all the wood I-joists with a simple jig (see the photo at right) as we take them off the pile and sort them by length and location. When I finish with the I-joists, I build any LVL headers and add joist hangers if they're needed.

After the prep work is done, I usually call in a crane to set all the steel beams that will carry the first floor and to spread all the pre-sorted stacks of joists and LVLs to their appropriate locations. I also move the sheathing to within 3 ft. or 4 ft. of the foundation so that I don't have to carry it any farther than necessary.

Square one end of each joist while sorting and stacking. Because I-joists are cut to approximate length at the lumberyard, it is easier just to square one end as you are sorting them.

Positioning joists. Position the square ends of each joist to the chalkline (the rim-joist line) and tack them into place along their 16-in. o.c. layout lines. Later, the 8d nail will act as a hinge when the joists are stood upright.

Spread Joists to the Layout Marks and Roll Them Upright

After placing the steel, I make sure that the layout on the beams matches what is on the walls. I check the layout by pulling a string from front to back to verify that the layout marks on the front and back walls intersect the marks on the beams. I also make sure that the beams are straight and flat, and make any necessary adjustments.

With one person on each side of the foundation, we quickly position the joists on their layout marks, with the square-cut end aligned on the rim-joist line snapped along the mudsill (see the photos above). Then we tack each joist in place with an 8d nail to keep it in place. It's easier to set the joists to the line first and then install the rim joist later. After tacking down all the joists, we prepare to cut the other end of the joists in place. We snap a chalkline that is 1⅜ in. from the outside of the mudsill, which becomes the cutline.

Cutting the joists to their finished length is as simple as running the saw along the chalkline using the I-joist cutoff guide (see the photo below). The scrap of wood lands in front, where it's available for use as a piece of blocking.

We position one person in the front and one in the back, and starting from one end, we stand all the joists and nail them in place (see the photo below right). The 8d nail that had held the joist in place now acts as a hinge for it. We usually can stand all of the joists for 40 lin. ft. of floor in about 10 minutes.

After aligning the joists, snap a chalkline and cut them to length in place. Beware of anchor bolts lurking below when making this cut.

You can do this alone, but it sure goes quicker with two. With one person at each end, stand the joists upright and put them on their layout marks. Drive one 10d nail through the flange on each side of the joist into the mudsill (or pony wall).

Don't Be Afraid to Hire a Crane

Many people associate cranes only with big commercial jobs, such as skyscrapers or shopping malls. But today cranes are commonly available for residential work, and anybody can hire one.

With a crane and one helper, I can set all the steel for a house and distribute stacks of pre-sorted materials to where they're needed. This process usually takes about 1½ hours ($180* here in Denver). This easily is cheaper than paying labor to move all that material, and we get to the framing faster.

The rim joist goes on after the joists are in place. The rim joists are cut and nailed to the mudsill every 8 in. with a 10d nail.

On each end of the I-joist and at the center beam, we put one 10d nail on each side of the joist through the flange into the mudsill. We keep the nails as far from the end of the joist as possible to avoid splitting the I-joist's flange. After standing the joists, we add the rim boards, cutting and nailing as we work our way around the house (see the photo at left). We put one 10d nail through the rim into the top and bottom flange of each I-joist.

Once the rim joist goes up, the last thing to do before sheathing is to add bearing blocks, also known as squash blocks (see the top right photo on p. 115). One person details the rim joist for bearing blocks, and another follows behind and nails them in place.

Bearing blocks are required anywhere that concentrated loads land on the joists, such as doorways or where a post supports a beam. We also put them at all inside corners, because 90 percent of the time this spot is a bearing point.

Lay the first row of sheathing plus two more sheets. Then move the rest of the stack onto the floor. It takes about 10 minutes to move 40 sheets; it's much quicker than having to climb up and down to get every sheet.

Sheathe your way over to where blocking is needed. Do not walk across unstable joists or work from a ladder below the floor. Sheathe over to the beam, then add the joist blocking.

Stack Sheathing on the Floor as Soon as Possible

We snap the line for the first course of sheathing 48½ in. from the outside edge of the rim joist. It's held back a little from the rim to account for any inconsistency in the rim joist.

Before we begin nailing the sheathing, we look for joists that may have been moved from the 16-in. o.c. layout. If our plywood joints are able to avoid them, sheathing will go much faster. After deciding on a starting point, we spread construction adhesive on top of the joists. Then we lay the first row and two sheets of the second row (see the bottom left photo on the facing page). This approach creates a little staging area where we can stack the rest of the sheathing.

We sheathe over to the steel beams in the center and add any bearing blocks and joist blocking when we get there (see the right photo on the facing page). Waiting until the floor is partially sheathed before installing blocks is a lot easier and safer than trying to balance on unbraced joists.

We cut all the blocks and spread them across the edge of the sheathing (next to the beam), starting at one end and grabbing them off the sheathing as we go. Layout marks for each joist on the plywood's edge keep the joists straight and plumb, and the spacing for blocking consistent. When we have a finished basement, we also add wall ties as we work our way across the floor, which keeps us from having to walk across unsupported joists.

When we get to a stair rough opening, we sheathe over it and brace the plywood seams. Not only is this approach safer, it also creates more usable floor space when we start framing walls. Before we stand any walls that surround the stair opening, we open it up again. If the hole is too large to sheathe over, we add a safety rail.

Lay as Many Full Sheets as Possible

As we sheathe, we lay as many full sheets as possible (making the fewest number of cuts). I've found that running the sheets long at the ends and cutting to a chalkline snapped along the rim (see the photo below) turns out a better product than measuring and cutting the pieces to fit individually.

I pull the chalkline in an extra ⅛ in. from the outside of the rim; this eliminates ever having to cut the rim line again. One person starts at a corner of the house and snaps all the rim lines; the other follows behind with the saw. The rim joist is first straightened and then nailed to the sheathing every 6 in.

Prices are from 2003.

Jim Anderson is a framing contractor in Littleton, Colorado.

Cut the sheathing in place. Run sheets long at the ends, snap a chalkline, and cut off the excess. This process is faster and turns out a better floor than cutting each piece to fit.

Framing and Sheathing Floors

■ BY RICK ARNOLD AND MIKE GUERTIN

We've finished backfilling the foundation, and the mudsills are level and square. Now the real fun begins: saws screaming, hammers humming, sawdust flying. But as anxious as we are to shift into high-gear production mode, we always approach the task of floor framing methodically and thoughtfully. With this strategy, everything goes together right the first time, and the reciprocating saw and the cat's paw stay in the toolbox where they belong.

A Good Framing Plan Streamlines Layout and Installation

Before we even think about getting our tools out, and usually before breaking ground, we start our floor on paper with a framing plan (see the illustration on the facing page). Most of the house plans we work from do not include a framing layout, so usually we create our own.

First, we choose the best starting point for the joist layout to minimize the number of joists and the subfloor waste. After looking at how the house is laid out (where the jogs are; how the roof trusses will be laid out; where bearing walls are; where toilets, tubs, and showers fall), we decide where to begin the layout. With the house featured in this chapter, the natural starting point was the front left corner because of the two adjoining sections where the joists changed direction. When in doubt, we usually pick the 90-degree corner that has the longest uninterrupted legs. All smaller sections are then blended into the larger layout.

On our plans we draw lines for each joist, header, in-floor beam, and any special framing details for the house. By using a different colored pencil for each joist length, we can use the plan for accurate material ordering later. When materials are delivered, the joist plan also enables us to direct the different-length 2×s to the appropriate areas of the floor with just a glance.

The Best Floor Layout Begins on Paper

Before any lumber is ordered, a detailed framing plan should be drawn. The framing plan exposes potential problem areas such as bearing walls, plumbing, or floor openings that might require special attention, and the color code indicates joist lengths for ordering and then precutting lumber when it arrives.

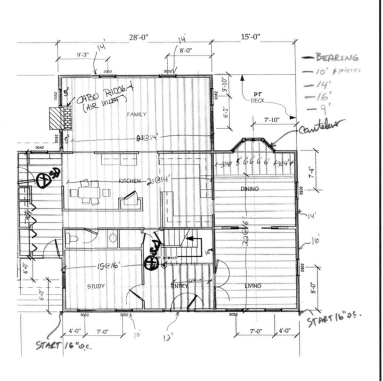

If you do the layout carefully and precut all the joists, the floor will go together quickly like a giant jigsaw puzzle.

We try to have a preliminary meeting (or at least a couple of quick faxes) with the plumber and heating contractor to identify any joists that might pose a problem with their systems. We can also alert each contractor if we see that both of them expect to fill the same joist bay. By moving a joist a couple of inches to one side or to the other, we sometimes can resolve competition for space. We generally try to avoid having a joist positioned directly below a wall above, and knowing which interior walls will contain vents, drains, or ductwork keeps us from placing a joist where it might have to be cut.

The big exception to that rule is where the house has bearing walls running parallel to the floor joists. In those cases we usually double or triple joists under the bearing wall to carry the weight. However, if a plumber or HVAC sub plans to use a bearing wall for drains or ducts, we identify the exact location of the wall on our plan and place the joists under the outside edges of the wall. Solid blocking is then installed between the joists every 2 ft. or so, leaving space for the systems to come through. We also double the joists beneath large tubs or whirlpools if the fixture is to sit in the middle of the joist span.

Floor Details Are Spelled Out in the Layout

Once the foundation is poured and backfilled, we take great care installing and adjusting mudsills, carrying beams, and basement bearing walls. As we prepare for the floor, everything is kept level and square, and the dimensions on the plans are matched exactly. The closer that we keep the tolerances at the floor-deck stage, the quicker and easier the rest of the house framing will proceed. Before beginning our layout, we string the carrying beams and bearing walls and brace them to keep them straight. These strings are left in place so that we can double-check the walls again after the joists are installed.

To ensure consistency, one crew member does all the joist layout. We begin our layout by marking any special features of the floor deck that interrupt common joist layout. In addition to the chimney and stair openings, the project featured in this chapter had a cantilevered section, in-floor beams, and two areas where the joists change direction.

These special details and measurements are marked with a lumber crayon on the sill plates to alert the crew that the standard layout has changed. Someone following the layout person can then precut the odd pieces, and the installation is easy and obvious.

Make Sure the "X" Is on the Right Side of the Line

With all the special features of the deck laid out, the next step is laying out the common joists on the sill plates that run perpendicular to the joists. This floor called for 2×10 floor joists laid out on 16-in. centers.

Starting from the end we determined on our plan, we make marks ¾ in. shy of each 16-in. symbol on the measuring tape. When the entire sill plate or beam is marked off, we go back and make a square line at each mark and draw an X forward of each line. The same procedure is repeated on the opposite side of the house starting at the same end.

Next, we run a line between our starting marks on the sill plates across any intermediate bearing walls or beams. This line gives us a reference point from which to lay out the tops of the bearing walls or the carrying beams (see the top left photo on the facing page). The uniform starting point helps to keep the joists in a straight line and makes it easy to lay down the subflooring later. Walls and beams are then measured and marked at 16-in. intervals from that point.

The main body of the floor featured in this chapter is 44 ft. deep. The span is broken into two 14-ft. sections and one 16-ft. section. The joists in the front section will be set ahead of our marks, as indicated

String keeps the layout uniform. The blue string in the foreground was stretched between the layout marks on the sills. Measurements on the carrying beams are then taken from that line.

Is the joist straight or crowned? A crew member sights each joist to determine the direction of the crown.

Mark the direction of the crown. An arrow is drawn on the board to indicate the top edge of the board.

by the Xs. The middle section will be set behind the mark, and the rear section will be set ahead of the mark like the front. The two outside sill plates get only one line to indicate the location for each joist. But on the two internal beams where joists from adjoining sections will overlap, we add additional lines indicating the outside edges of the joists. (Because we toenail the overlapping joists to the beam after both joists are in place, a single layout line would be hidden beneath the joists.)

Careful Attention Is Paid to Crowns

While one crew member works on the layout, another sorts and crowns the joist stock (see the top right photo above). We use only kiln-dried lumber for floor joists. Kiln-dried lumber is less likely than green lumber to

shrink or to change shape over time. And because kiln-dried lumber is preshrunk, we don't get problems usually associated with green stock, such as drywall cracks, cracked tile, and doors and windows that bind.

Every piece of sawn dimensional lumber has a crown, or a natural curve it takes on after it is cut from a log. We look at each floor joist and mark the direction of its crown with an arrow (see the bottom photo above). Those with excessive crowns (more than ¼ in. in 8 ft.) are set aside to be cut into window or door headers later.

Rim joist goes on first. Before any common joists are installed, a rim joist or band joist is nailed to the outer edge of the mudsills. While one crew member assembles the rim, another transfers the layout up from the sills to the inside face of the rim joist.

with hardwood receive B-joists, and C-joists with the biggest crowns are saved for floor areas under carpet or to be cut into headers.

The Rim Is Installed First

The rim joist or band joist is toenailed to the outer perimeter of the floor on top of the sill plates or to the top plates of exterior walls (see the top photo at left). We use 16d nails every 12 in. The rim joists that run perpendicular to the layout prevent the floor joists from rotating. Rim joists that run parallel to the layout close off the floor area along its outside edge. We also install band joists at the interior transition points where joists change direction. Here, they serve as a break point for the edges of the sheathing as well.

We select straight stock for the rim joists so that the crowns don't leave a space between the rim and the plate. If such a space is left, the rim will eventually settle under the weight of the house and cause problems later. When straight stock is scarce, we make a sawcut near the middle of the rim joist about two-thirds of the way across the board (see the bottom photo at left). The cut is made in the direction of the crown and lets us fasten the joist down all the way to the plate.

We use the rim joists as in-floor headers over window and door openings in framed walls wherever possible. In-floor headers let us skip the traditional headers and jack studs for openings in exterior walls that run parallel to the joist direction. As long as the rim doesn't break over the opening, a single rim joist can carry the wall weight above short openings. For wider spans such as over a sliding door, we double up the rim over the opening.

This method uses a little less lumber framing, and more important, it increases the thermal efficiency of the wall. On this house we eliminated 24 jack studs (or 3 ft. of solid wood in the walls) and 36 ft. of header stock. All this space can now be insulated.

Taking the crown out of a rim joist. If a rim joist has a severe crown, a relief cut is made that allows the joist to be drawn all the way down to the sill or plate.

When we get a unit of joist stock that has many boards with crowns of more than ¼ in., we grade each joist with an A, B, or C designation. Without this extra effort, we could end up with large differences between adjacent joists, creating a washboard effect in the floor and making it difficult to install the tongue-and-groove sheathing. In those cases, the straightest A-joists are used as rim joists and beneath tiled areas such as kitchens and baths. Floor sections that will be covered

Precut pieces for framing floor openings. Using measurements written on the sills during layout, kits containing all the framing members for floor openings such as stairs or chimneys are cut and labeled ahead of time.

Rolling the joists into place. After the joists are laid in flat, a crew member rolls them onto their layout marks and nails them to the rim.

Common Joists Are Rolled into Place

We used to take care to cut the rim-joist stock to break exactly on the center of a floor joist. But because the structural wall sheathing extends down to cover and secure the joint, there isn't any real benefit in doing so. We do check the end of every rim joist to make sure that it's square and trim it if it's not. Square ends are especially critical at the corners to maintain the exact dimensions of the floor deck. Once all the rim joists are in place, we use a framing square or a triangular rafter square to square up all the layout lines from the sills or basement wall plates onto the inside of the rims.

The crew member who crowns the floor joists also checks the end that will butt against the rim joist for square. At the same time, framing members for floor openings are cut from the measurements written on the sill plates and then grouped into kits (see the left photo above). For example, the kit for this house's chimney consisted of short rim-joist pieces, headers, and cripple joists. When the kit is finished and each piece is clearly marked, it is neatly stacked outside the foundation close to where it will be installed. In areas where joists change direction, the joists have to be cut to length to fit between two rim joists. After being cut, these joists are also stacked near where they will be installed.

We usually assign one crew member to assemble and install the kits for rough floor openings, and the rest of the crew installs the common joists. We first lay all the joists flat on the sill plates and across the carrying beams with all the crowns facing in the same direction (see the photo on p. 121).

Just Say No to Bridging

Here in the Northeast, where construction quality is often judged by how much wood you can pack into a house frame, omitting bridging is a controversial choice. Unless we exceed the 6:1 (depth-to-thickness) ratio in our joist material (2×12) where the CABO code requires blocking or bridging between joists, we almost never install it.

To the best of our knowledge, bridging has never been proved to add strength to a floor, but it is almost certain to add squeaks. In the past we tried gluing in solid blocks, and for several years we installed steel bridging as an alternative to blocking. We'd install the steel bridging tighter than a guitar string only to return a year later and find it had loosened and was causing squeaks. Even though we use kiln-dried material, the seasonal changes in humidity cause the joists to shrink and swell enough to render any type of bridging worthless.

We can just about guarantee a squeak-free floor in our homes unless our clients or an architect insists on blocking or bridging. In those cases, we are inevitably called back a year later to fix floor squeaks. The remedy for the squeaks usually involves removing any solid blocks or bridging that didn't have to be removed when the plumbers and HVAC installers did their work.

Instead, ¾-in. tongue-and-groove structural sheathing glued and nailed to the joists is effective at transferring loads to adjacent joists, which is what blocking and bridging are supposed to do. As extra insurance, we install a continuous 1×3 strap nailed to the underside of the joists down the center of the span in the basement to keep the joists from twisting. If the ceiling is to be finished, such as above a living space, we install 1×3 strapping 16 in. o.c. across the whole ceiling.

Squash blocks carry loads from above. Two-by blocks called squash blocks cut slightly longer than the height of the joist help to transfer loads directly to the sill plate or carrying beam. Here, the squash blocks are installed under header-bearing jacks for a sliding door above.

Now we can walk along the outside of the foundation or on top of the plate rolling the joists into place and nailing them to the rim (see the right photo on p. 125).

If a joist is shorter in height than the rim, we lift and nail it flush with the top of the rim. We go back later and shim under all the short joists. After a joist is nailed through the rim with four or five 16d nails, we drive three toenails through the joist and into the mudsill or top plate. At this point, however, we don't nail the joists at the beams or bearing walls.

After all the joists are nailed in place, we recheck the strings that we set up earlier to straighten all the interior carrying beams and walls as well as any exterior framed walls. When we're satisfied that everything is straight, we walk the beams and nail the overlapping joists to each other again, flushing the tops and shimming under short joists. The overlapping joists are fastened to each other with four or five nails driven at an angle so that the nail points don't stick out the other side. The joists are now set on the outside lines we drew earlier and toenailed to the beams or wall plates with four nails.

According to code (CABO 502.4.1, 1995), each joist must bear a minimum of 1½ in. where it sits on a carrying beam, and there must be a positive connection at the joist laps. There are three basic ways to make an approved connection between joists that overlap. The most common way to connect opposing joists is by overlapping them a minimum of 3 in. Another method is using either a wooden block or a steel connector plate as a splice across the joist joint. The third method is letting the subfloor sheathing span across the intersection of the joists by a minimum of 3 in.

We never use solid blocking between joists over the beams or bearing walls to transfer loads. Instead, we frame all our walls so that the studs line up directly over the joists. Wherever we have concentrated loads falling on a joist from a wall above, such as jacks carrying a load-bearing header, we install squash blocks, a technique we borrowed from our engineered-I-joist experience. Squash blocks are 2× blocks cut slightly longer than the height of the joist. They are installed on end beside the joist to help transfer loads to the sill plate or carrying beam (see the photo on the facing page). Usually, we install squash blocks after the floor is sheathed, unless we can pinpoint bearing points before. Since we began using squash blocks in conventional floor decks, we've virtually eliminated drywall cracks around door and window openings.

Don't Skimp on the Glue for the Sheathing

When all the joists are fastened in place, we double-check all the floor-deck dimensions and take diagonal measurements to make sure the deck is square before we start

Tweaking the rim joist. Before the sheathing goes on, a string is run along the top edge of all the rims that run perpendicular to the joists. The rim is tapped in or out and checked with a square until it is perfectly straight.

installing the subfloor sheathing. If the rim joists were installed with square ends at the corners, our measurements are usually close. If the diagonal measurement is off more than ¼ in., we tweak the rim in or out to make the adjustment.

Next, we run strings along all the rim joists running perpendicular to joist runs. We tap the top of the rim in or out as needed and shim if necessary to get the rim joist perfectly straight (see the photo above). Squaring the joist ends helps to keep these adjustments to a minimum. Rims running parallel to the joists will be straightened later after the sheathing is installed.

To begin sheathing the deck, we measure 4 ft. in at both ends from our starting edge, usually the front of the house. We snap a line and start pumping adhesive onto the joists. From there we snap lines every 47½ in. for a glue guide for each row of sheathing. (We use a 47½-in. measurement

Start spreading the glue. A generous bead of construction adhesive is spread on each joist and on the plywood edge.

A sledgehammer snugs the sheathing into place. A 2× block protects the grooves in the edges of the sheathing as it is tapped into place with a sledgehammer.

because the tongue and the groove cost us ½ in. for each row of sheathing.) The snapped line tells us where to stop the glue for each row to keep glue off our tapes and to keep the joists beyond the sheathing safe to step on.

The first set of sheets is set with the tongues on the rim joist so that we don't ruin them when we bang the sheets into place on successive rows. The first course of sheathing is nailed off completely with 8d ring-shank nails so that it doesn't drift when we drive the next set of sheets in place. We don't glue or nail the edges of the sheets along the rim joist running parallel to the joist direction so that it can be straightened later.

Spreading glue on the joists is an often-overlooked operation. However, we take our gluing seriously. We probably go through many more tubes than most crews, but we believe it's worth the extra labor and material.

Each joist gets a generous bead of glue, and the section of joist where two panels meet gets a bead along both edges of the joist (see the top photo at left). We also run a bead of glue down the groove before installing the next row of sheets. The glued

A single nail keeps the joist on the layout. After the sheet has been tacked at the corners, a tape is hooked on the joist tacked to the sheathing, and the other joists are moved until they fall into position. A single nail is then driven to hold the joist on the layout.

Added support at the overlap. The joist layout changes wherever the joists overlap at an interior bearing wall or a carrying beam leaving part of the edge of the sheathing unsupported. A 2× block is nailed in to hold the unsupported edge (photo above), and the layout is clearly marked for accurate nailing (photo right).

tongue-and-groove seams are much stiffer and squeak-free than those left dry. Plus, when glued properly, the sheathing functions as a vapor barrier, provided that all utility penetrations are carefully sealed.

Framing crews usually just flop the sheets of sheathing down haphazardly onto the joists and slide them over into position. In the process, the glue is smeared and rendered useless, and the joists become a sticky, slippery mess. Instead, we try to lay each sheet down as close to where it is supposed to go and as carefully and gently as possible, which keeps the glue bead where it belongs and keeps the work area neat and safe.

As each sheet is laid down, it is tapped against the adjacent sheets with a 2× block and a sledgehammer (see the middle photo on p. 128). The OSB floor sheathing we use lies flat, and the tongues slide easily into the grooves. When plywood is specified, it's usually necessary to have an extra crew member stand on the seam to flatten the sheet. We adjust the joist that falls under the end of that sheet so that half or about ¾ in. of the joist is left exposed. The outermost corner of the sheathing is then nailed to secure the joist in position. After each course of sheathing is tacked in place in this manner and before the next course is started, we hook a tape onto any of the secured joists and measure, adjust, and nail the rest of the joists at their proper 16-in. o.c. position with a single nail at the edge of the sheathing (see the bottom photo on p. 128).

We stagger the butt joints between sheets 4 ft. with each successive course. When the layout approaches an area where the joists overlap, we install 2× blocking to support the end of the sheet as needed (see the top photo on p. 129). After we've tacked the whole field of sheets in place, we snap lines to indicate joist locations, taking care to shift our lines where the joists overlap or change direction (see the bottom photo on p. 129). One crew member then finishes all the nailing so that he can keep track of what's been nailed.

Straightening the final edge. When all the sheathing has been installed, a line is snapped 1½ in. from the ends. The rim joist is then moved in or out until it lines up with the end of the tape.

Tying Up Loose Ends

Whenever our sheathing runs by an opening such as the stair chase or the opening for the chimney, we either let a small section of sheathing overhang or leave a small uncovered area to begin the next sheet at the edge of the opening. With the bulk of the floor sheathed, we now turn to these details, trimming and filling in as needed.

Because the sheets of sheathing are 47½ in. wide, we end up with a 5-in. void at the end of our 44-ft. house. Rather than sacrificing several sheets of sheathing for their tongues, we cut up scrap pieces of sheathing and use them as fillers. The unsupported joint is not a concern because it will be covered by the 2×6 walls above.

It's rare to have a floor this wide. Most of the houses we do are less than 30 ft. wide, so the sheathing shortfall is usually 3 in. or less. If we don't have enough scrap sheathing to use as a filler strip, we use 1×3 strapping or rip a 1×6 ledger instead, which is cheap and easy to use.

The last step is straightening the rim joists that were left unnailed during the sheathing operation (see the photo at left). Measuring in 1½ in. from the corners at the ends of the rim, we snap a reference line on top of the sheathing. We now move the rim joist in or out every 4 ft. or so until the distance from our line to the outside edge of the rim measures 1½ in., and we drive a nail at that point. After the edge is tacked straight, we nail it off and trim off any excess sheathing.

The second-story floor deck is built pretty much the same way. However, one step that we take just before lifting the bearing walls that will support the next floor is laying out for the joists on top of the top plate, which is quicker than doing the layout from staging. Once the walls are up, we start the process of installing rims and joists all over again.

Fine Homebuilding *contributing editors and authors of* Precision Framing, *(The Taunton Press, Inc.),* **Rick Arnold** *and* **Mike Guertin** *are builders and residential consultants in North Kingstown and East Greenwich, Rhode Island.*

Built-Up Center Beams

■ BY RICK ARNOLD AND MIKE GUERTIN

We looked at a basement remodel recently. But before we got to the basement, the owner was showing us large cracks in the tile floor in the kitchen and entry. She hadn't noticed the drywall cracks forming in some doorways. We'd seen these symptoms before, and in the basement we found that the center beam had been built of green lumber and had shrunk almost ¾ in. So before the remodel could begin, we had to jack up that center beam so that it could do its job: Hold up the house.

Built-Up Beam Basics

In the simplest terms, a built-up center beam provides a straight, level surface that supports the floor joists between the walls of the foundation. Like most, the beam we installed for this project was even with the mudsills and was carried by columns set on footings at regular intervals.

A built-up beam is made of several layers of lumber nailed together and set on edge. The beam for this project was made of dimensional lumber, but laminated veneer lumber can also be used (see the sidebar on p. 132). The number of layers and the size of the lumber are determined by the load that

Keep beam flat during preassembly. For a straight beam, sections of the beam are preassembled on the ground and must be kept flat, with the tops of the boards kept flush. Just a few nails join the layers at this point.

Although we install a lot of dimensional-lumber center beams, laminated veneer lumber may be a better choice when you want to reduce the number of support columns and simplify the installation. LVL boards span greater distances than similar-size dimensional lumber and are much less prone to shrinking. They also come in longer lengths.

LVL boards are 1¾ in. thick and from 7 in. to more than 18 in. deep. Layers of an LVL beam are joined together with nails or bolts just as with a dimensional-lumber beam. You can usually purchase and handle lengths that reach the full distance between beam pockets, or at least halfway, minimizing butt joints.

LVL beams are dimensionally stable, so they can be set level with the mudsills without allowing for shrinkage. Also, LVL tends to be straighter and have no crown, so LVL beams don't take much tweaking to get them true.

On the downside, LVL beams are heavier than dimensional lumber, so plan your crew accordingly. A 36-ft. LVL beam that's 9½ in. high is more than two people can handle safely. LVL is also much more dense than regular lumber, which makes nailing the layers together more difficult. Pneumatic nailers don't always drive nails in completely. We find that clamping the members together before shooting the nails helps, and any nails that aren't sunk completely can be sent home with a hammer.

Long, straight, and strong comes at a price. Expect an LVL beam to cost twice as much as a dimensional-lumber beam.

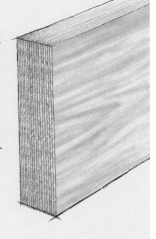

the beam has to carry, the species of lumber, and the span between support columns.

Most center beams fit into recesses in the foundation called beam pockets. The simplest center beam spans from one side of the foundation to the other. With larger or more complex designs, there may be several beams, and some beams may span only a portion of the basement width.

Sketch the Beam before You Start

Before we order materials for a new house, we sketch a beam plan that shows the numbers and lengths of the boards in the beam. On the sketch we mark the centerpoint of each support column and the measurement between those points as well as to the inside edges of the foundation.

Most center beams are longer than the longest available stock lumber, so we plan for butt joints in each layer. With an engineer's approval, butt joints can fall between columns, but rather than take chances, we locate all butt joints over the columns. These joints should be staggered between layers. On the sketch, we label each beam layer by number and the boards in each layer by letter to keep things organized on the job site.

Before we assemble the beam, we measure from the corners of the foundation and mark the exact center of the beam on the mudsills above the beam pockets (see the top left photo on the facing page). Working from the center, we draw the edges of each beam layer on the mudsill.

We also set up A-frame scaffolding to support the beam temporarily as we set it in

Center beam goes here. After the crew measures from the corner of the foundation, the exact location of the center beam and all its individual layers are marked on the mudsills.

Up, over, and into the pocket. Preassembled beam sections are easy for two crew members to lift. Here, crew members slide one end of the beam into the foundation pocket and rest the other end on the A-frame scaffolding.

Wedges keep the beam standing up. Temporary 2× wedges that are placed in the pocket keep the beam from moving while assembly continues.

place (see the top right photo above). The tops of the A-frames put the beam close to its final height, and staging planks on the lower cross bars put us in a good position for assembling and positioning the beam.

Keep the Beam Flat during Assembly

At this point, we cut the pieces of the beam to length. Each board is given its piece-and-layer label, and the direction of the crown is marked. The boards are then spread out in their approximate location on the ground inside the foundation.

It's much easier to assemble part of the beam and lift it into position on the A-frames rather than build it from scratch in position. And although we've seen it done, we never preassemble the entire beam and try to lift it into place, which is a dangerous proposition regardless of the size of your crew. So starting at one end of the beam, we line up the second layer on top of the first. We keep the tops of the boards flush as they're nailed together (see the photo on p. 131).

Nails staggered every few feet are enough to hold the two layers together at this point. We try to keep the beam as flat as possible during this process. Any waves built into the beam as it's tacked together can be hard to take out later. For long center beams (more than 40 ft.), we assemble two or three of these two-layer sections, orienting the crowns in the same direction. The first of the assembled sections is then lifted onto

Almost a beam. The last board in the first layer completes the bridge between the foundation pockets. At this point, the first two layers are still only tacked together.

the A-frames and slid into the beam pocket (see the top right photo on p. 133). Blocks of 2× wedge the beam section upright temporarily (see the bottom photo on p. 133). Next we add preassembled sections or additional pieces until the first two layers are complete from pocket to pocket (see the photo above).

Brace the Beam to Keep It Straight

Before adding the rest of the layers, we brace the beam straight so that no curves are built in (see the photo at right). We stretch a string the length of the partial beam, spacing the string from the beam with two short 1×3 blocks. A third block is used as a gauge.

Before nailing on the braces, we make sure the beam hasn't sagged. If it has, we adjust the A-frames until the beam is approximately level again. Then we extend

One piece at a time. After the first two layers are braced straight and nailed off, each of the final layers is added one board at a time. After the entire layer is in place, the crew goes back and nails it off.

an adjustable 2×4 brace from the mudsills across the top of the beam at each support-column location. These handy braces, available through concrete-form supply houses, consist of a turnbuckle that is then attached to a 2×4. The adjustable ends of the braces are nailed 2 in. in from the edge of the mudsill so that they don't interfere with the rim joist and floor joists.

Keeping the beam roughly straight, we nail the other end of each brace to the beam. Then one crew member fine-tunes the brace on the turnbuckle end while another gauges the beam with a block. Besides keeping the beam straight, the braces keep it from rolling over while we add the final layers. When the two layers have been braced straight, we fasten them together permanently with rows of 12d or 16d nails every 12 in. to 16 in. The nails are driven at an angle, so they don't poke through the other side.

Add Final Layers One at a Time

The next layer can now be added to the beam with the crown up and the top flush with the rest of the beam (see the bottom photo on the facing page). When that layer is tacked in place, we make sure the beam is still straight before nailing it off.

If the beam has a fourth layer, it is added the same way. As we add successive layers, we're careful to keep the joints staggered and to install each board according to our sketch.

Set the Beam a Little High

Dimensional lumber always shrinks. A two-plate mudsill that's 3 in. thick can be expected to shrink ⅛ in. to ¼ in. over the first year or two as the house dries out. Even though we try to build our center beams of kiln-dried lumber, a beam can shrink up to ⅝ in.

Lifting a Beam without Jacks

Here is a site-built alternative for raising a center beam without screw jacks. First, lay down an 8-ft. 2×6 in the basement on a flattened area of ground that runs perpendicular to the beam just to the side of the column location. Cut two 2×6s about 5 in. longer than the distance from the 2×6 on the ground to the bottom of the beam, and nail the two 2×6s together at the tip of one end with the other ends spread apart. Slide the nailed ends under the beam, and place the loose ends on the flat 2×6 to form an A-frame (see the photo at right).

Tack the top of the A to the underside of the beam, and tap the legs inward equally to make the beam rise. If more than one lift is being used, adjust each a little at a time and equally along the length of the beam until the beam is at the correct height. To keep the legs from kicking out, nail on 2× blocks to back up each leg.

depending on its moisture content when it's installed. To compensate for this shrinkage, we install the beam ¼ in. to ⅜ in. above the top of the mudsills.

To level the beam, we first stretch a series of strings between the mudsills perpendicular to the beam, one string over each column location. We always use strong twine that can be pulled tight without sagging. Just as with straightening the beam, spacer blocks are placed under the strings at each end.

Screw jacks that raise the beam are then placed about 1 ft. from each column location, snugged up to the bottom of the beam. If screw jacks aren't available, there is a site-built alternative (see the sidebar on p. 135).

With the screw jacks in place, the A-frames can be removed. We adjust the jacks until the beam is ¼ in. to ⅜ in. higher than the mudsills, nudging each jack a little at a time to bring the beam slowly up to position (see the photo above).

Permanent Shims in Beam Pockets

Once the beam is at the correct height at each of the column locations, we move to the beam pockets and install permanent shims to lift the ends of the beam to the same height. Shims should be made of dimensionally stable material that won't crush or rot.

The thickness of the space below the beam affects the choice of shim material. Spaces ½ in. or less are best filled with steel plates. Several plates can be stacked to fill the void, or steel plates can be used in combination with plywood (see the top left photo on the facing page). Softwood shims, such as cedar shingles, should never be used to shim a center beam.

Pressure-treated lumber shims should never be used with the grain flat. In this position, they can shrink and allow the beam to settle. Instead, we install these shims with the grain in a vertical position

Beam ends rest on permanent shims. Shims that support the ends of the beam can't rot or be crushed. Steel plates and plywood can be used for small spaces (photo above left), and lumber shims with the grain in a vertical position are used to fill larger spaces (photo above right).

(see the top right photo above). These shims work best to fill spaces 3 in. or more.

We also replace the temporary blocks on the sides of the beam pockets with pressure-treated blocks. To make sure that the beam stays on its layout marks while the floor is being framed, we nail steel straps in an X between the beam and the mudsill (see the photo at right). We usually make the X out of the steel strapping from the lumber delivery.

Recycled steel strapping keeps the beam in place. An X made of steel strapping from the lumber delivery anchors the beam to the mudsills while the floor is being assembled.

Columns: Now or Later?

Most of the houses we build call for 3½-in.-dia. concrete-filled steel columns (also known as Lally columns) as permanent supports. The columns can be installed at this point (see the sidebar on pp. 138–139), but because there is no weight on the beam yet, we sometimes wait until the floor system has been installed. When they're supporting the weight of the floor, they're less likely to be knocked out of place accidentally.

But don't wait too long. Once load-bearing walls and the weight of a second floor have been added, it may not be possible to lift the beam without more powerful jacks.

Fine Homebuilding *contributing editors and authors of* Precision Framing, *(The Taunton Press, Inc.)* **Rick Arnold** *and* **Mike Guertin** *are builders and residential consultants in North Kingstown and East Greenwich, Rhode Island.*

Installing the Support Columns

Before the columns can go in, we first make sure the beam is still level. If our strings have been stretched for more than a day, we retighten them to take out any sag. Then each post location is marked on the bottom of the beam, and the distance from the concrete footing to the bottom of the beam is measured at each post location.

If the layers on the bottom of the beam aren't flush, we chisel off a flat spot for the top column plate. We also take off any high spots on the footings for a smooth surface. The measurements are written on the beam. To obtain the actual cutting length, we subtract the thickness of the top and bottom plates if one or the other isn't already welded to the column, and mark that length on the column.

It's best to cut longer columns first so that if one is cut too short, it can be recut and substituted for one of the shorter columns. A large pipe cutter is the fastest and easiest way to cut a column, but it's an expensive tool to own and not always available to rent. An alternative is using a metal-cutting blade in a reciprocating saw or circular saw. But all cutting options begin with an accurate cutline around the circumference of the column.

The easiest way to draw the cutline is to wrap a large piece of paper around the column (see photo 1), keeping an edge of the paper lined up on the mark and then matching the edge of the paper to itself as it wraps around the column. A pencil line follows the edge of the paper.

We gently cut through the metal skin of the column, following our line (see photo 2). When we're most of the way through the steel, a light tap with a hammer breaks off the waste piece. If the concrete core of the column breaks off beyond the cutline, a few hammer taps on the concrete chips off the excess.

If the top plate hasn't been welded to the column, we fasten it to the underside of the beam. To let each column slip in more easily, we raise the beam about ⅛ in. with the screw jack. The column is then slid into place and set on the baseplate. Next we roughly plumb the column and lower the jack until the column starts to bear the weight of the beam. The bottom of the column can then be tapped into place while it's plumbed with a level (see photo 3). When the column is set, we make index marks around the baseplate just in case it gets bumped out of place during construction (see photo 4).

If the slab hasn't been poured, the base of the column can be secured to the footing with a couple of masonry nails. If the slab has already been poured over the footings, we install lag shields into the concrete and then bolt down the baseplate of the column. As a final step, the column is welded to the top and bottom plates. We usually make a continuous weld around the ends of the column with a MIG welder. The small tabs on the plates should not be trusted to keep the column in place.

1

Paper makes the perfect cutline. A sheet of paper with the edge set at the measurement and then wrapped around itself creates a continuous line for cutting around the column.

That's a big pipe cutter. A large pipe cutter is the easiest and fastest way to cut a steel column to length. A circular saw or reciprocating saw with a metal blade can also be used.

2

3

Perfectly plumb. After the support column and plates are inserted under the beam, one crew member taps the column into plumb while another keeps his eye on the level.

4

Column base belongs here. After the column has been plumbed, the position of the column base is marked on the footing in case the column is bumped before the basement slab locks it into place.

Installing Floor Trusses

■ BY BRIAN COLBERT

As a framing contractor, I think speed and efficiency are my most important considerations. I'm not wasteful, but I don't care as much about the cost of the material as the speed with which I can put it together. Although parallel-chord floor trusses are more expensive than conventional floor joists, my five-man crew can set the floor trusses for a 2,800-sq.-ft. house and nail down the plywood subfloor all in one day, for labor savings of 25 percent over 2× joist installation. I believe those labor savings more than offset the extra expense. What's more, I believe we end up with a stronger floor.

There are several reasons that floor trusses can be installed faster. Trusses are engineered to exact lengths, so no blocking and cutting are required. As a bonus, their webbing—the diagonal and vertical 2× material inside the truss—provides more than enough room for wiring, plumbing, and mechanicals, which means there should be no ill-planned cutting, boring, or notching to repair later.

Trusses are not without their weaknesses—chiefly that they can't be cut or modified in any way—but their strength means fewer bearing walls, which also makes the job go faster. Trusses are relatively light and easy to handle, and they require no furring out for drywall. And you should have no callbacks because of squeaky or bouncy floors.

Also, trusses are more stable over long spans than conventional stick framing (see the photo on the facing page), and their open-web design makes adding wiring, plumbing, and mechanicals simpler than wood I-joists. Also, the 3½-in. wide top chord of a floor truss provides plenty of nailing surface for the decking.

Trusses Are Engineered for Each Job

To order trusses, the general contractor gives the manufacturer's representative a complete set of plans. The representative then hands the plans off to an engineer, who designs the trusses based on the spans and the loads for my application. For instance, if a homeowner wanted to put a whirlpool in the master bath, the engineer might specify two trusses directly beneath the tub or drop the spacing down from 24 in. o.c. to 16 in. o.c. or even 12 in. o.c.

All floor-truss installations begin with this plan. On the plan, each truss is numbered or lettered to correspond to its location. Trusses

Parallel-chord floor trusses are more stable and span greater distances than stick framing. Speedier setup offsets their higher cost.

arrive at the job site marked with numbers or letters that correspond to their place in the plan. When they're delivered, however, the trusses often are out of order. If we have a crane to lift them, we're in pretty good shape. But if we're lifting them by hand, it's a real pain when the trusses we need to set first are buried beneath the ones we'll set last.

I try to be on site when the trusses are unloaded. If I'm there, I can save my crew time by making sure that the trusses are unloaded in an organized manner. Usually, trusses are loaded on the truck in a way to make the most economical use of space.

As soon as the trusses arrive, I check each one and report any damage to the fabricator and to the engineer or architect overseeing the job: No one should try to repair a damaged truss. Also, the trusses should be set on blocks and stored as flat as possible to prevent them from deforming over uneven ground. If they're to be stored for an extended period, it's best to cover them with a tarp.

Even more important is the need to measure all the trusses against the plan as soon as they're delivered (see the left photo on p. 142). I make sure we have the right spans, the right number of trusses, and the right types. I've installed trusses on a dozen houses, and it's not uncommon for some trusses to be either too long or too short. Sometimes, this situation is the fault of the fabricator. Usually, though, such mistakes result from a lack of communication between the architect and the fabricator. Because it takes an average of one to two weeks for trusses to be delivered—twice that time in some areas or during peak times—it's not unusual for changes to be made in the design of the house during the interim.

In any event, an order of trusses often contains a small percentage of trusses that are either too long or too short—there may even be extra trusses. Even if I conclude that

TIP

If I know the order in which I want the trusses unloaded, I can call the manufacturer's representative and ask to have the trusses loaded in a certain order. This type of sorting takes a lot of time at the plant, though, and not all fabricators are willing to go to the extra trouble.

Make sure you get what you ordered. After inspecting each truss for damage, the author measures the trusses to make sure they're built according to the framing plan.

Trusses Are Fragile until They're Installed

It's not necessary to lift the trusses in place with a crane, but the $80* per hour we pay seems like a bargain, given the speed with which we get them in place. If you're using a crane, however, it is critical that trusses be lifted correctly. When set in place and properly fastened, parallel-chord trusses are wondrously strong. Until they are properly set, however, they are fragile. A sling should hold the truss one-quarter the length of the truss from each end (see the photo below). We run heavy rebar or steel pipe through the webbing at those points and hook the crane's chains around the metal.

At this point, a good crane operator can be invaluable. If the operator knows his trade, he also knows where to set bundles of trusses in a way that makes the job easier for

the fabricator is at fault, I'm still looking at a possible two-week delay before new trusses are delivered. That's a long time when I've got a crew standing by to frame the rest of the house.

On the job shown here, one section of trusses was 1½ in. too long. We had a choice of having the manufacturer take them back and deliver the correct lengths—and waiting two weeks or so for them—or working with what we had. We decided to set the trusses as delivered and to fur out the rear foundation wall with 2×s so that the siding would fit flush.

Also, several of the trusses delivered to the job site were too short, which meant we had to frame kneewalls to fill the gap. That error was the result of a change in the plans that went uncommunicated to the truss manufacturer.

The sling distributes the weight in fourths. With one-quarter of the length of the truss remaining on each end, the crane sling connects to heavy pieces of rebar slipped through the webbing. A worker steadies the trusses and turns the bundle to ensure the trusses all face the right way.

the crew. We were lucky on this job to have Charles Ellrey as our crane operator. Charles's 30 years of experience was a big help.

Even with a crane, the crew still has to move some trusses by hand. It's natural to want to turn trusses on their side when lifting them by hand. That's fine if they are under 15 ft. or so in length. But if they're longer than that, they can fold or snake out of form. What's more, if long trusses are temporarily set on their sides, they can sag enough to wedge themselves between their sills or plates. This is less of a problem if the bearing surfaces are 2×8 or greater.

Remember These Few Safety Precautions

Trusses are held together by galvanized-metal plates, which are engineered for a particular load and span. The plates are punched through to form teeth, which are pressed into the intersections of wood webbing and chords, holding them together. However, the teeth in these plates are razor sharp and can easily rip a chunk out of an ungloved palm.

Gloves are a nuisance, and most builders prefer to work without them. But when you're lifting and moving trusses, it's a good idea to wear leather gloves to protect vulnerable hands. Occasionally, we encounter loose plates, which can really deliver a nasty cut.

The main safety concern with floor trusses, however, occurs when they are set on the plates or bearing walls but not yet fastened or braced. If a bundle of eight or ten trusses is set down by the crane, we nail a brace across the tops of them to make them stable (see the photo below). I've seen more than one carpenter thrown down to the foundation floor by walking on unbraced

At this point, a good crane operator can be invaluable. If the operator knows his trade, he also knows where to set bundles of trusses in a way that makes the job easier for the crew.

Temporary braces prevent rollover. This bundle of trusses in the foreground may look stable, but without a temporary brace, they can roll over and cause serious injury. In the background, workers nail a temporary brace across trusses that have already been nailed down. Conventional joists were used in one section of the project because of the short span and light loads needed for that area.

Ladder trusses go along outside walls. Ladder trusses are made specifically for outside walls, where their vertical webs make an easier target for sheathing and siding nails than diagonal webs would be.

trusses. Although it would seem that a bundle of trusses resting on 3½-in. wide chords would be steady, they can easily roll over. Even when floor trusses are nailed down, it's still not a good idea to walk over them unless you've nailed a temporary brace across them.

Start Setting the Trusses at an Outside Wall

Ladder-type trusses, which have all-vertical webs, are installed at outside walls (see the photo at left). It's much easier to nail wall sheathing and siding onto vertical webs than onto diagonal webs. We set these farthest trusses first so that we're not walking over other trusses any more than necessary.

For safety reasons, it's necessary to brace the first truss. We usually nail a 2×4 diagonally to the mudsill and nail the top end of the 2×4 into that first truss. The brace should keep the truss from rolling over as we work over and around it.

A permanent ledger connects the trusses. Once a 16-ft. row of trusses is nailed down, 2×4 ledgers can be fastened to the notches provided at the ends of the trusses.

After setting a row of trusses—say the first third in a row—we nail a temporary brace across the top chords. This brace provides a reasonably stable surface for walking. Once we've set a 16-ft. row of trusses, it's time to nail the permanent 2×4 ledgers into the pockets provided at the ends of the trusses (see the photo below).

Occasionally—based on the design-load conditions—one or more strongback braces may be required. These braces are usually 2×6s that run vertically inside the open webs from truss to truss, resting against the bottom chord. A strongback brace usually is required when the size or lumber grade of the truss has been exceeded. If a strongback is required, those instructions will accompany the truss delivery.

Orientation of Trusses Is a Critical Factor

Bottom bearing and top bearing are the two main types of floor trusses. Top-bearing trusses have extended top chords, which rest on the sill or plate, allowing the rest of the truss to hang below. These trusses are helpful when the height of the building is a concern. Bottom-bearing trusses—which we show here—rest on the bottom chord like a conventional joist. Either way, it's important to make sure that the trusses are turned the right way when, end for end, they are set on a bearing surface. Usually, there's a number or mark painted on one end of each truss to signify which end goes which way.

One reason the trusses must be oriented correctly is so that the webbing aligns. The webbing usually contains squared chase openings for heating and ventilation ducts, so it's critical that these squares be aligned before the trusses are nailed in (see the left photo on the facing page).

There is another good reason why it's important to orient floor trusses as planned. Trusses often are required to make multiple spans from plate to plate over one or more

This square chase doesn't align. Several trusses were inadvertently installed backward and had to be reversed so that the square duct chase would align.

Truss design provides bearing points. Load-bearing points are identified on the truss by a spray-painted mark or a tag and must fall over the bearing surface, such as this partition wall. When two trusses meet over a wall, one truss will have a pocket, and the other truss will have an overlapping top chord that fits into the pocket.

bearing walls. In order for them to work in this fashion, trusses are designed with special load-bearing points that have been built into them. If one truss is to meet another truss over an interior bearing wall, one truss will have a ledger pocket, and the other truss will have an overlapping top chord, which fits into the pocket for nailing the two trusses together (see the right photo above). Truss manufacturers typically identify the load-bearing points on these interlocking trusses with a spray-painted mark or a tag at the points that fall over the walls.

When Things Don't Go According to Plan on the Job Site

The *Metal Plate Connected Wood Truss Handbook* (Wood Truss Council of America; 608-274-4849) warns that "drilling holes and notching may cause immediate deflection of the truss and possibly contribute to the collapse of the entire structure. If drilling or notching seems necessary, a truss-design engineer must be contacted to review the consequences of such field modifications."

Clearly, builders are not to do anything to trusses except install them. Yet there are times when it's necessary to modify a truss—if ever so slightly—without waiting for an engineer's okay. The most common modification occurs when the truss gets set over an anchor bolt. You either drill a hole in the bottom chord of the truss to fit it over the bolt, or you move the truss one way or the other and put the rest of the joists off layout. We prefer to drill out the bottom chord carefully to accept the bolt.

Another common problem I've faced is when the bearing point of the truss does not line up with the bearing surface of an interior wall. Our truss maker staples a piece of orange paper to each bearing point. On this job, the maker fabricated the trusses and located the bearing points based strictly on the plans provided. However, the plumber failed to rough in the plumbing on one section exactly as called for. So when we built the load-bearing wall that would hide those drains and supply lines, the wall was off

The camber in these bottom chords isn't a problem. Sometimes the bottom chords of long trusses contain a distinct camber that prevents their contact with the bearing walls below (see photo at right). The author levers the truss over the layout marks (see photo at far right) and drives a nail into the bearing surface on both sides of the truss to hold it in place.

Framing Openings with Floor Trusses

Conventional framing around a fireplace, chimney, or other opening requires headers that are either hung from joists or nailed directly to the joists. Truss framing provides several options for framed openings. I'll briefly describe the two most prevalent types.

DOUBLE TRUSSES

The most common method of framed openings that we use calls for double trusses on all sides of the opening (see the top drawing). This method is similar to conventional framing, except that only trusses are used. The longer spans on the sides of the opening are doubled, and the smaller double trusses that enclose the front and back of the opening are hung off those longer trusses. Special double-wide truss hangers are usually provided by the truss manufacturer.

POCKET GIRDER

The other common method involves specially made trusses that include a pocket for a girder to header off one or two sides of the opening. These could require double trusses perpendicular to the header. If double trusses are used, the girder, or header, slides through both trusses (see the bottom drawing). If single trusses frame in the two sides of the opening, the girder or header—which can be a glulam, microlam, laminated-veneer lumber, or built-up 2×s—slides through the pockets of as many trusses as it takes to distribute the load adequately.

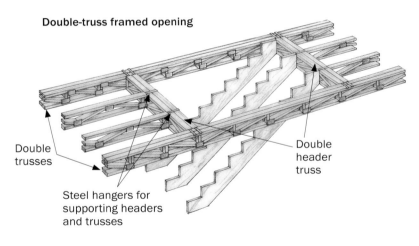

Double-truss framed opening

Double trusses

Steel hangers for supporting headers and trusses

Double header truss

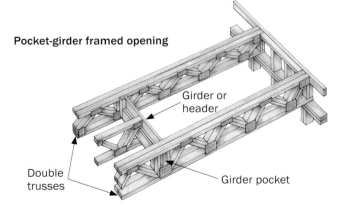

Pocket-girder framed opening

Girder or header

Double trusses

Girder pocket

from the plan about 2½ in., which meant the bearing points of our trusses were off. If we had known in time that the walls were that far off, we could have communicated that to the fabricator, who would have relocated the bearing points.

In such a case, I recommend getting an engineer's opinion. He may specify additional bracing or plywood gussets between the wall and the truss.

Another, although minor, problem that often occurs is when the bottom chords of trusses that are longer than 10 ft. or 15 ft. contain a distinct camber that prevents their contact with the bearing walls below (see the left photo on the facing page). Rather than forcing the chord down to the bearing surface and nailing it in, we pry the truss over the layout marks on the bearing wall using a 2×4 lever. Then we drive a nail into the top plate of the bearing wall on both sides of the truss, thus holding it over the mark (see the right photo on the facing page). That way, the weight of the house will settle the truss in the right spot.

There's No Special Trick to Nailing Trusses

In general, there are no special requirements for nailing floor trusses. We usually toenail two nails through the bottom metal plate into the bearing surface. We also add two nails straight through the bottom chord into the top plate or sill. This type of fastening is sufficient for ordinary live and dead loads. The plywood subfloor and exterior sheathing form a box with the trusses that keep them in place. However, when you get into wind loads, seismic loads, and headered openings, trusses often need special anchors or hangers (see the sidebar on the facing page). A number of companies make both anchors and hangers required for special truss installations (see "Sources" at right). However, whenever special hardware is required for installation, the truss manufacturer usually supplies it.

Pry a wayward chord into place, then nail. Although the bottom chords are nailed according to layout, the top chords often need to be forced into plumb. While one man pries the chord into place, another nails the plywood to it.

Nailing down the subfloor over floor trusses is a breeze. Almost anybody can hit the broad side of a 2×4 with a nail. One thing to remember, though, is that with particularly long trusses the top chord may be out of plumb with the bottom chord even after the truss is set. In other words, the top chord may have bowed to one side, which can be a problem when it's time to nail down the subfloor.

We solve this problem by using a level to plumb the top chord on the first truss. Then we brace that truss, set the first sheet of plywood on it and nail the plywood to it. From that reference point, we pencil layout marks on the plywood to show us where the top chords should be. We use a 2×4 to lever the top chords into position. One person pries the truss into position (see the photo above) while another nails the plywood to the top chord. Rim joists or band joists are unnecessary when using parallel-chord floor trusses. The plywood wall sheathing does the job instead.

Prices are from 1998.

Brian Colbert *of Colbert Construction Co, Inc., is a builder in Boone and Banner Elk, North Carolina.*

Sources

The following companies make anchors and hangers required for special truss installations.

KC Metals
408-436-8754

Semco
800-737-7327

Simpson Strong-Tie
800-999-5099

Framing Floors with I-Joists

■ BY RICK ARNOLD AND MIKE GUERTIN

Most people wouldn't think twice about spending a couple hundred dollars on a computer upgrade that might be obsolete shortly. But the same people would labor for weeks deciding whether to spend a like amount upgrading the quality of a new home by using wood I-joists instead of standard lumber for floors.

I-joists are a bit scary the first time you handle them. Short 3-ft. or 4-ft. pieces seem sturdy enough, but 30-footers flop around like al dente linguini. I-joists take a little more care to install than dimensional lumber, but when installed properly, they produce a stronger, flatter, and stabler floor deck than one framed with the best kiln-dried dimensional lumber.

I-Joists Can Be Part of an Engineered Floor System

I-joists are one component of the engineered-wood floor systems manufactured by numerous companies. There are small differences between floor systems from one manufacturer to the next, but we haven't found

those differences critical enough to affect either performance or installation.

Besides the I-joists, two other components combine to form engineered floor systems. The first is the main beam or girder that carries the joists. In a conventional deck, the main beam usually consists of layers of dimensional lumber such as 2×10s sandwiched together. In an engineered floor system, LVLs or PSLs are used instead. LVLs and PSLs span greater distances and are more stable than beams made with conventional lumber.

I-joists can be installed over dimensional-lumber carrying beams, although manufacturers of engineered floor systems caution against it. If, however, the center beam is installed in-floor or in the same plane as the I-joists, with the I-joists hung from the sides of the beam, a PSL or LVL beam is a must.

The other element of an engineered floor system is tongue-and-groove plywood or OSB structural panels—with the APA Sturd-I-Floor® rating—that goes over the joists (see the top photo on p. 150). Here again, conventional structural panels can be used over I-joists, but it goes against manufacturers' recommendations. Also, putting

Whether you're building a new house or remodeling an old one, engineered lumber can give you squeak-free floors with fewer callbacks.

I-joists are the heart of an engineered floor system. Plywood or OSB sheathing is nailed and glued to I-joists for a strong, stable, and squeak-free floor system.

Larger I-joists span greater distances. I-joists come in a variety of sizes, and the larger the webs and flanges, the more load they can carry and the greater distance they can span.

Factory-made rim stock is load bearing. Instead of nonstructural ripped-down sheathing, 1-in.-thick manufactured stock the same height as the I-joists creates a load-bearing rim.

Sturd-I-Floor sheathing down using construction adhesive and the proper nailing schedule actually increases the allowable span of the joists.

As the name implies, I-joists are shaped like the letter "I." The vertical sections are called webs, and the horizontals are the flanges. I-joists come in a variety of sizes, and those with larger webs and flanges can carry larger loads and span greater distances (see the middle photo at left). Manufacturers assign a series number to each size indicating the strength and spanning capabilities of the joist. But be careful when comparing companies' I-joists. Series numbers are not standard from one manufacturer to another.

An Engineer Can Help to Size Up Your I-Joist Needs

When converting conventional floor framing to I-joists, the manufacturers or distributors provide an engineer free of charge to help with the conversion. If we are drawing the plans ourselves, we bring the engineer in during the design phase. It's a lot easier to integrate I-joists into a house at the design stage, and I-joists offer more design freedom because of their longer clear spans.

We can usually lay out and size the I-joists for most floor systems just by following the literature and charts provided by the manufacturer. Knowing that the lowest series of any I-joist is an upgrade from any common framing-lumber species of the same height, it's usually just a matter of replacing dimensional lumber with the similar-height engineered member (a 2×10 floor joist would be replaced with a 9½-in. I-joist). But because defects don't exist in I-joists, we don't have to order extra stock to replace boards with bad crowns, giant knots, or splits.

We install the floor systems ourselves, so we try to design them as simply as possible. Here, the engineer's input is invaluable. Adding an extra beam, a joist, or a hanger can save labor in the long run. An engineer also helps us to eliminate unnecessary components. We've found that the more complicated the house, the easier it is to frame the floors with engineered lumber.

Until recently, we had to use ripped pieces of ¾-in. plywood, OSB, or an I-joist for the rim joists that run perpendicular to the floor framing. But now stock rim-joist material is available for I-joist floor systems (see the bottom photo on the facing page). Rim stock is made from 1-in. or 1¼-in.

material similar to OSB and manufactured to the same height as the I-joists. Manufactured rim stock has a high load capacity and in most instances eliminates the need for squash blocks, or solid blocking between the ends of the joists that helps to transfer the outside wall loads to the sills. Manufactured rim stock is available in 12-ft. to 16-ft. lengths that are straight and strong.

It's a lot easier to integrate I-joists into a house at the design stage, and I-joists offer more design freedom because of their longer clear spans.

Why Use I-Joists?

Wood I-joist systems have a number of advantages over dimensional lumber. The first and most obvious is that they weigh about half as much as kiln-dried lumber. A worker can easily carry a 12-in. high 40-ft. joist (see the photo at right), although for safety's sake—and to keep from damaging the I-joists—we try to have two crew members handle long joists. The I-joist height tolerance is within ¹⁄₁₆ in. We feel lucky if our kiln-dried joists vary less than ¼ in.; ⅜ in. is more the norm.

Another advantage to I-joists is their dimensional stability. Even the best kiln-dried lumber shrinks or swells after installation, which can cause a wide variety of problems, from squeaking floors to cracks and nail pops in drywall. I-joists don't suffer from these size fluctuations.

I-joists are perfectly straight with no crowns, so there is no need to eyeball every member as we have to do before installing regular lumber. Also, unlike dimensional lumber, I-joists don't check, split, warp, or twist. Electricians and plumbers like I-joists because holes for pipes and wires can be larger, and it's easier to cut through I-joist webs than through solid lumber. The biggest drawback to I-joists is that they cost more than dimensional lumber. We got quotes for the house in the photos from several wood I-joist dealers who carry national name

brands as well as the regional varieties. Then we priced the floor made out conventional materials. The price for engineered lumber averaged about 15 percent higher, but your decision on using I-joists shouldn't be based solely on price.

One factor that can give perspective to the price difference is that longer lengths of dimensional lumber cost more per foot than shorter lengths. But because I-joists are a manufactured product, lengths over 40 ft. are available at no additional cost per foot. And then there's the intangible cost factor of all the technical and engineering support as well as performance guarantees that come with I-joists, something that doesn't come with a lift of 2×12s.

I-joist manufacturers claim that the higher cost for their product is offset by the labor savings realized during comparable installations and the callbacks that are avoided using dimensionally stable I-joists. We agree that I-joists make a superior floor deck with fewer callbacks, but we haven't found any appreciable labor savings in the installation of I-joists compared with conventional lumber.

I-Joists Should Be Handled with Care

Handle with care. I-joists are more fragile than regular lumber and should be handled with care. Damage such as a split flange renders the I-joist unusable.

Handling I-joists on a job site is the biggest concern for us. They are much more prone to damage than solid lumber. We never let the delivery driver dump a load of I-joists. The flanges can be damaged if the I-joists are dropped onto a rough or rocky job-site surface (see the top photo at left). A ½-in. chip out of a 2×12 won't affect its performance, but a similar chip could be disastrous for an I-joist.

We've seen flanges pop off the webs when they're dumped from a delivery truck. Don't try to reattach an I-joist flange. The repaired joist will fail eventually. The best, most careful way to unload I-joists is by hand or with a boom truck.

After the joists are unloaded, stack them on a flat surface or block them up to keep them from sagging. If left in a twisted position on the ground for too long, they may take on an undesirable shape. The integrity of the joist probably won't be affected, but it will be much harder to handle and set in place. Stacked flat, I-joists sort of lock into each other for a strong and compact pile (see the bottom photo at left).

I-joists also need to be carefully supported for cutting. Joists longer than 24 ft. should be handled by two workers to avoid damage. We stack the joists on blocks set flat on the ground or on sawhorses spaced no more that 10 ft. apart. Although I-joists are rigid in an upright position, I-joists set on their sides will sag if they're not supported, which makes measuring lengths more difficult. Also, be careful not to saw through the flanges of the next I-joist in the pile when you're cutting that top joist. A sawcut in a flange ruins the I-joist.

Stacked I-joists lock together. When stored at the job site, I-joists should be blocked up off the ground at least every 10 ft. to prevent sagging.

A site-built jig speeds up the cutting. Because of I-joists' irregular shape, accurate and straight cuts require a jig. A square scrap piece of sheathing that fits between the flanges supports the saw table that rides along a guide screwed to one side of the scrap.

I-Joists Have to Be Cut to Length

When laying out the sills for I-joists, we have to make adjustments to keep the joists centered on the layout because the I-joist flanges are usually wider than 2× joists. The plan we receive from the manufacturer or distributor notes any special items or conditions that we have to pay attention to, such as squash blocks, web stiffeners, backer blocking, and hanger locations. We transfer any of these special notes to the plates or beams in red crayon so that the joist installers won't miss important structural details.

After the joist layout is complete, we toenail the rim-joist stock to the sills just as we would with a conventionally framed deck. When laying out I-joists and setting rims, we pay close attention to the minimum bearing surface required by the wide-flange I-joists in the joist plan. Some I-joists may need more bearing surface on the sills or wall plates than is left once the rims are set. If this is the case, we can install wider sills or increase the thickness of the wall, but usually we just skip the rim joist for the time being and run the joists all the way to the outside edge of the sill plate or beam. We'll go back later and block between the joists with rim material.

All ends should be checked for square. I-joist ends should be checked and cut square before they are cut to length. Out-of-square ends are a common occurrence.

After the rims are installed, we carefully measure and cut each joist to length. Don't trust the factory ends of I-joists. They rarely come through square (see the bottom photo above). Cutting I-joists is easy and accurate with a simple jig that can be made on site (see the top photo above). We write the measurements on the plate and keep the cut

Squash blocks help to transfer point loads. Two-by-four blocks cut slightly longer than the height of the joist transfer point loads directly to the sill.

Web stiffeners add strength to the I-joist web. When I-joists are asked to carry extra loads in specific places, web stiffeners are added. Sheathing the same thickness as the flange overhang is nailed to both sides of the web to supply additional support.

One detail called for in almost every I-joist floor is squash blocking, which helps to transfer the weight from load-bearing walls directly to mudsills, center beams, or wall plates (see the left photo). I-joist manufacturers specify that squash blocks be cut $\frac{1}{16}$ in. higher than I-joists to ensure that the squash blocks take most of the load. Squash blocks are made from 2× stock usually mounted flat against the top and bottom flanges of the I-joist.

Typical locations for squash blocking are beneath jack studs that carry large headers for sliders or French doors or beneath beam-carrying posts. Sometimes they're called for on top of the center beam alongside the I-joists to support a center bearing wall. Squash blocks are also used beneath exterior walls where non-load-bearing stock has been used to close in the rim.

Among other details specific to I-joists are web stiffeners, which are pieces of OSB or plywood thick enough to fill in the width of the flange beyond the web (see the bottom photo at left). The framing guide says that web stiffeners should be cut about $\frac{1}{8}$ in. less than the distance between top and bottom flanges, probably so that they can be inserted without driving the web from the flange. They are called for in areas of concentrated load to strengthen the web and to strengthen the webs of some of the taller I-joists. Also, if an I-joist is going into a hanger that isn't tall enough to catch the top flange, web stiffeners are added to keep the I-joist stationary in the hanger. Manufacturers don't give exact dimensions, but web stiffeners are usually only as long as the width of the 2× load they're helping to transfer. Web stiffeners are applied in tandem and clinch-nailed through the web and opposite stiffener.

I-joist headers and doubled I-joists that carry headers are built a little differently than solid-lumber headers. I-joist headers need a 2× filler block between the webs of the mated joists running the length of the header. On I-joists that are doubled for carrying headers, filler blocks are installed

I-joists in order so that they can be dropped into place without confusion.

Long I-joists should be carried by two people to make sure they're not damaged from flopping around. It is critical to position the I-joists directly on their layout marks on center beams or bearing walls before nailing the ends. Long I-joists can sometimes vibrate off their intermediate layout points several inches while being handled. We always drive 6d nails beside the flange into the beam on both sides of each I-joist to keep the joist properly positioned until the ends are nailed. I-joists are fastened to the sill with 8d nails driven through the flanges.

I-Joist Floors Require Special Framing Details

Concentrated load points in I-joist floors require details not usually found in conventionally framed floors. These areas are indicated on the plans from the manufacturer or distributor. Also, every I-joist package comes with a floor-framing-detail guide that depicts nearly every framing peculiarity you may encounter.

between the webs at the locations of all intersecting joist hangers. In these locations the filler blocks usually extend a couple of feet to both sides of the hanger location.

We select the stock size of the 2× filler block to match closely the width between the top and bottom flanges of the I-joists. You don't need to rip the blocks, just use stock with the closest nominal width. Nails fastening two I-joists together should be driven through both webs, through the blocks between and then clinched over. We never nail the I-joists together through the flanges. Nails can split LVL layers in the flange and cause the I-joists to fail.

Use Only Specially Designed Joist Hangers

Backer blocks are yet another framing detail used in I-joist floors. These blocks are made of OSB or plywood and look a lot like web stiffeners, only they are usually much longer. Although web stiffeners help to increase the compressive strength of the I-joist, backer blocks are used to back up joist-hanger locations or to back up the web in areas where the I-joist has to be nailed to a framing member (see the top photo at right).

When hangers are called for, we use only hangers specifically designed for I-joists. Most I-joist hangers are top-flange hangers, which means that the top of the hanger has a horizontal tab that attaches to the top of the I-joist flange or beam (see the bottom photo at right). Occasionally, U-shaped hangers without the top flange are specified, but these also are designed especially for I-joists, with wider seats to accommodate the flange.

Regardless of type, all hangers hung on I-joists need to be installed with backer blocks that extend 9 in. to 12 in. on both sides of the hanger. Top-flange hangers require backer blocks under the top flange on both sides of the joist to support the flange and to keep it from rotating under load.

Backer blocks protect the I-joist web. Plywood blocks on both sides of the I-joist keep the web from being damaged when the I-joists are being nailed to framing.

Special joist hangers are used for I-joists. Top-flange joist hangers have horizontal tabs that wrap over the flanges of adjoining I-joists or LVL beams.

Backer blocks for U-shaped hangers should rest against the bottom flange and be high enough to accommodate the entire hanger. Never use the bottom flange of an I-joist to support a load. Loading the bottom flange may make it separate from the web.

As we do with web stiffeners, we make our backer blocks out of stock the same thickness as the width of the flanges beyond the web of the I-joist. Backer blocks should always be installed on both sides of the web and clinch-nailed the same as web stiffeners. We also run backer blocking to fill the web spaces that are

A bit of insurance will bring silent floors. A squirt of construction adhesive in the bottom of the joist hanger helps to prevent floor squeaks.

Plywood blocking for anchoring interior walls. Lengths of plywood are tacked to adjacent I-joists to attach nonbearing interior partitions that run parallel to the joists.

left exposed at stair openings and in open foyers to make the drywaller's job easier.

We've found that if these so-called squeak-proof floor systems are going to squeak, they're most likely to do it around joist hangers. To help ward off these annoying noises, we make sure all I-joist hangers are nailed off properly. Then, for added insurance, we squirt a little construction adhesive into the hanger seat before dropping the joists into place (see the photo above left).

Structural Panels Complete the Floor

We've developed one detail that makes it easy to attach the top plates of walls running parallel to I-joists overhead (see the photo above right). Before we install the floor sheathing, we locate the joist bays that fall above parallel nonbearing walls. We rip scrap pieces of sheathing to fit between the webs of adjacent I-joists, and then we tack them down to the top side of the bottom flanges every 16 in. to 24 in. with 6d nails. We always air-drive these nails as opposed to pounding them in with a hammer, which could weaken the web-to-flange connection. Two-by blocks are then attached to the sheathing blocks for a great nailing surface to secure top plates.

As mentioned earlier, wood I-joists are designed to be just part of an engineered floor system working together with the subfloor structural panels to complete the system. After the I-joist installation is complete, we apply a layer of ¾-in. tongue-and-groove subfloor that is glued to the tops of the I-joists with construction adhesive. We also put adhesive between the tongues and grooves of the sheets themselves. Ring-shank nails are used to attach the subfloor.

As you begin to frame with I-joists, you'll likely encounter many unusual framing details. We were once faced with something that can be described only as an outside loaded flying cantilever. We sketched the problem and fired off a fax to the distributor's engineers. After phone conversations and faxes back and forth, we came up with a detail we could use.

Fine Homebuilding *contributing editors and authors of* Precision Framing, *(The Taunton Press, Inc.)*, **Rick Arnold** *and* **Mike Guertin** *are builders and residential consultants in North Kingstown and East Greenwich, Rhode Island.*

Supporting a Cantilevered Bay

■ BY MIKE GUERTIN

hen a client wants to add curb appeal to a new home, I dip into my Mr. Potato Head® bag of tricks: A distinctive window here, a reverse gable there, fancy trim details, an entry portico or a porch—and voilà! It's enough to make an architect cringe.

One of the best-selling upgrades is an angled bump-out or bay. It adds a few square feet, creates a distinctive room inside, and dresses up the home's exterior. Although I

could just install a bay window for light and effect, I find the floor-to-ceiling bay more appealing as well as competitive in cost.

But a bay is only as strong as the floor that it's built on. Here I'm going to concentrate on the proper techniques for framing the cantilevered floor that supports a bay. For this project, the bay was 8 ft. wide and extended 2 ft. from the house. The sides of the bay were set at 45 degrees, but they could have been set at any angle.

Plates before joists.
Before the cantilevered-bay joists are cut and installed, the wall plates are cut and laid out. At this point, all measurements are checked, and the exact width of the bay is established.

Cantilevered Joists Save Foundation Work

Cantilevering the bay keeps down the cost, about $400* less than an angled foundation. It's also easiest to frame one of these bays when the joists run parallel to the floor framing. In this scenario, the common joists are just lengthened to form the bay joists, eliminating the need for headers and hangers. But I wasn't so lucky on this project. The floor joists of this bay ran perpendicular to the main joists (see the photo below).

The cantilever wouldn't be carrying any loads but the bay itself, so I followed the two-thirds in, one-third out cantilever rule of thumb. With a 2-ft. cantilever, the bay joists would be anchored to a tripled floor joist 4 ft. in from the outside of the house.

But I waited to add the second and third joists until just before sheathing the deck. Having only one common joist allowed me to nail through it to attach the bay joists initially. The bay joists follow the 16-in.-o.c. layout regardless of exactly where the bay is placed, so first I put in all the 4-ft. joists that fell on each side of the bay area.

Cut and Lay Out the Bay Plates First

Before I laid out the exact location of the bay, I cut and laid out the top and bottom plates for the bay walls (see the photo on p. 157). Although this step may seem a bit premature, I always want to be certain that the windows will fit and that I'll still have room inside and out for the trim. The plates also help me to figure the length and cut for each joist.

Bay joists hang off the main joist. When the bay joists run perpendicular, they are nailed to a main joist. After the cantilevered joists are attached, the main joist is tripled and joist hangers are installed.

A little basic math and a calculator gave me the plate lengths. With the bay cantilevering 2 ft. and the walls at 45 degrees, I needed to come in 2 ft. from each side for the bay's front plate. With 22½-degree angles on each end (half of 45 degrees), I cut the plate for the bay's front wall at 4 ft. from long point to long point.

With some help from Pythagoras, I cut the side plates, again with 22½-degree angles on each end and with the outside face measuring 33¹⁵⁄₁₆ in. long point to short point (short point because the adjoining wall plate is also cut at 22½ degrees to form the inside corner). With the plates laid out on a flat surface, I marked the rough opening for the window centered on the 4-ft. plate.

To get the width of the trim (exterior and interior) to match on both sides of the bay's outside corners, I make sure that the distance is the same from the corners to the edge of the rough openings for all three windows. After the rough openings are marked out, I also make sure that I have enough space left (at least 1 in.) for the inside-corner trim.

Center the Bay on the Interior

I usually center a bay on the room inside. In this case, that threw it slightly off-center on the exterior, but the difference wouldn't be noticeable. I marked the location of the 8-ft. opening on top of and on the outside face of the sill plate. (On this house, the sill plate is actually the top plate of a framed wall for a walk-out basement.)

Next, I marked the outside corners of the bay on the sill, showing me which joists would cantilever the full distance. The house's rim joists were then run to the locations of the first cantilevered joists inside the 8-ft. layout marks rather than being mitered into the bay's rim joists. The extended rim joists are nailed square to the

Setting the overhang. With a 2-ft. cantilever, the longest joists overhang 22½ in. The joists are then tacked in place.

Before I laid out the exact location of the bay, I cut and laid out the top and bottom plates for the bay walls. Although this step may seem a bit premature, I always want to be certain that the windows will fit and that I'll still have room inside and out for the trim.

Bay starts here. After the bay joists are tacked in place, the outside edge of the bay is carried up from the face of the sill onto the rim joist.

cantilevered joists to hold them straight and plumb, and in turn, they provide a more solid place to secure and plumb the bay's rim joists where they meet the house wall.

The cantilevered bay joists were put in next. The joists that fell in the middle 4-ft. section of the bay cantilevered by 22½ in. (2 ft. less the thickness of the rim; see the top photo on p. 159). These joists were tacked to the sill and nailed to the common joist. The joists that fell on the angled sidewalls were left a little long and cut to exact length later.

When all the joists were nailed in, I ran a string to straighten the main rim joist and then drew square lines up from the lines I'd made earlier on the sills, indicating the outside edges of the bay (see the bottom photo on p. 159). The rim joist for the outer wall of the bay was then cut and nailed in, left long to be cut to exact length later (see the photo at left).

Outer rim joist is left long. Rim-joist stock is nailed to the cantilevered joists and left long until cutlines are transferred from the plates.

Precut wall plates provide the shape of the bay. The bay plates are laid on top of floor joists and rim joists that were run long and needed to be cut.

Use Plates to Mark the Joist Cuts

With the joists in place, I next set the bay's wall plates in position over the cantilevered joists and rim (see the bottom photo on the facing page). At the outside corners of the plates, I squared down a cutline indicating where to miter the bay's rim joist (see the photo at right).

A line was also drawn along the outside edge of the sidewall plates onto the tops of the joists that were left long. This line is the perimeter of the bay, so holding a 2× block inside the line and drawing a second line allowed for the rim joist and gave me the actual cutline. After squaring down the cutlines, the joists were trimmed at 45-degree bevels (see the photo below). If the amount of waste is more than about a foot, I rough-cut the length so that the cutoff isn't heavy and unwieldy. Finally, I cut, fit, and installed the angled rim joists (see the top photo on p. 162).

With all the floor framing for the bay complete, I tripled the main joist that the bay hung from and installed two joist hangers

Cutlines are squared down from the plates. Lines are extended down from the corners of the plates for the rim-joist cuts, and the lengths of cantilevered side joists are marked 1½ in. from the outside edge of the plate.

Cut down where they stood. The cantilevered side joists and the outside rim joist are cut to length in place.

Interlocking joint connection. The main rim joist on the house was run to the first cantilevered joist and nailed, keeping both joists plumb and square and providing a solid landing place for the angled rim joist.

Wait—those joist hangers are upside down! To prevent uplift, joist hangers are nailed on top of the bay joists as well as on the bottom, where they support the weight of the floor.

on each of the cantilevered joists, one right side up and the other upside down (see the middle photo at left). The theory is that the upside-down hanger prevents the joist from lifting from the weight of the cantilever. Some framing crews block between the cantilevered joists at the sill plate, but I prefer to leave the space open to slide in insulation later.

With floor framing done, I finished sheathing the main deck, extending the sheathing over the cantilever (see the photo below). I left the walls and roof of the bay until the entire house was framed. The house walls give me a nice surface for attaching the bay roof, and a roomy exterior deck wrapped around the bay, disguising the fact that it was sitting on a sturdy cantilevered floor framing.

Prices are from 2000.

Mike Guertin is a builder, a remodeling contractor, and a contributing editor to Fine Homebuilding from East Greenwich, Rhode Island. His website is www.mikeguertin.com.

Sheathing completes the floor framing. With the bay joists on regular centers, the deck sheathing can be cut and extended to include the bay without changing the sheathing pattern.

6 Ways to Stiffen a Bouncy Floor

■ BY MIKE GUERTIN AND DAVID GRANDPRÉ

If you haven't fallen into the basement of your house already, don't worry; your bouncy floor is probably not an indication of a disaster waiting to happen. Floor deflection is common in older homes because the floor joists often are smaller or are spaced farther apart than the joists in modern homes.

Of course, new homes also can have bouncy floors if the joists are approaching the maximum spanning distance for the weight they are supporting. Long-span joists may meet design criteria and the building code, yet still feel uncomfortable.

A well-designed wood floor feels stiff as you walk on it but still gives slightly under foot, absorbing some of the impact of your steps. Too much bounce, though, can make the china cabinet wobble. You can shore up floor joists and reduce the bounce in a number of ways, but the six methods outlined here represent a mix of common and not-so-common solutions. The best choice depends on access to the joists, obstructions in the floor system, or current remodeling plans; one technique or a combination may be your most practical solution.

It's important to make these improvements carefully. If existing joists have been weakened due to rot or insect damage, glue and fasteners won't hold well, and your work may be ineffective. Loose blocking or an underfastened subfloor won't bring any benefit, so take extra time and care during installation. Also, you can use jacks to relieve the load on joists while the work is being done. Jacks improve the effectiveness of your floor-stiffening work.

1. Built-Up Beams Are Rock Solid but Reduce Headroom

This solution works best in crawlspaces where you aren't too concerned about limiting headroom or cluttering the space with columns. If you don't mind the obstructions or loss of headroom, though, beams and columns certainly can be added in basements, too.

The important thing to remember when adding a support beam is also to add proper footings to support each column. In most instances, a 2-ft.-square, 1-ft.-deep footing provides adequate support. However, when you're installing LVL or steel beams with wider column spacing, larger footings may be necessary to support the load.

Steel or LVL beams can be used in lieu of dimensional lumber. Their added strength will allow for wider column spacing, but larger footings may be required to carry the load.

Steel beam

Laminated-veneer lumber

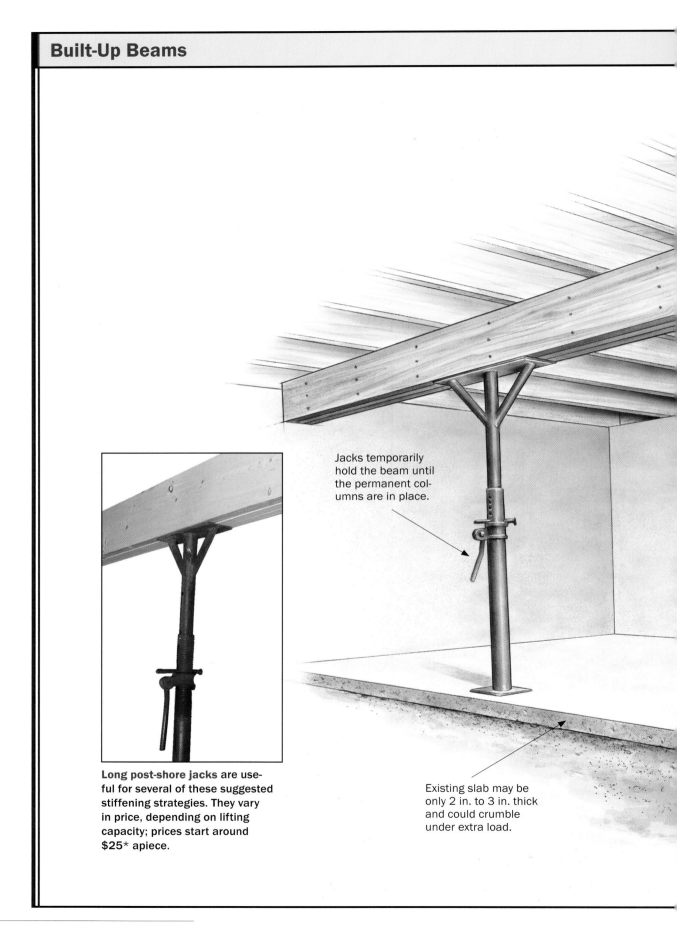

Jacks temporarily hold the beam until the permanent columns are in place.

Long post-shore jacks are useful for several of these suggested stiffening strategies. They vary in price, depending on lifting capacity; prices start around $25* apiece.

Existing slab may be only 2 in. to 3 in. thick and could crumble under extra load.

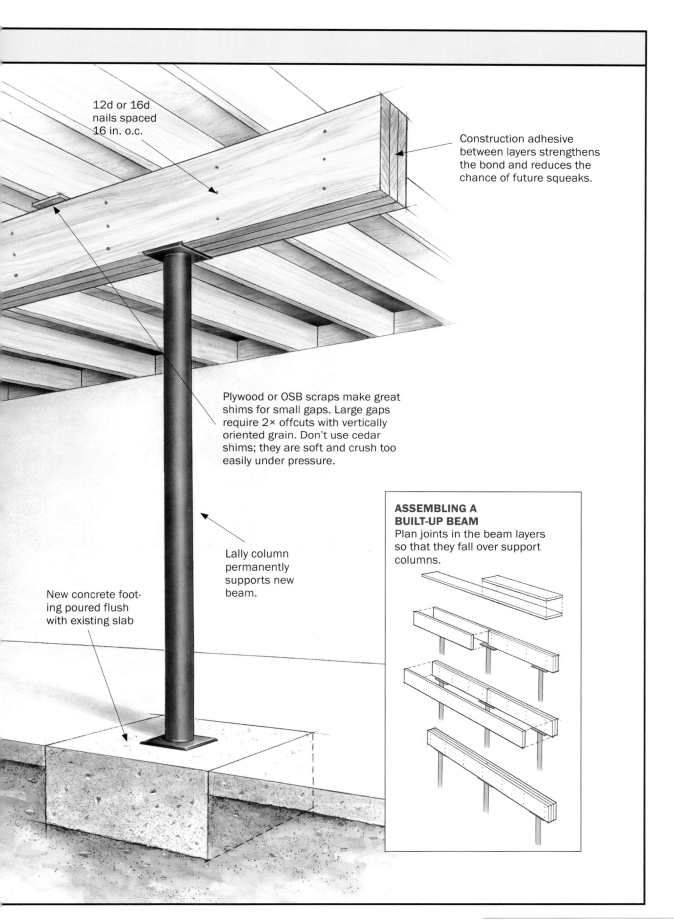

12d or 16d nails spaced 16 in. o.c.

Construction adhesive between layers strengthens the bond and reduces the chance of future squeaks.

Plywood or OSB scraps make great shims for small gaps. Large gaps require 2× offcuts with vertically oriented grain. Don't use cedar shims; they are soft and crush too easily under pressure.

Lally column permanently supports new beam.

New concrete footing poured flush with existing slab

ASSEMBLING A BUILT-UP BEAM
Plan joints in the beam layers so that they fall over support columns.

The beam size depends on the load and span of the beam between columns. Also, keep in mind that the closer you space the footings and columns, the more rigid the new beam will be and the stiffer the floor will feel.

To make a new footing, cut the slab, dig out the earth beneath, and pour concrete flush with the top of the slab. Next, snap a chalkline across the underside of the joists in the middle of the span to help align the new beam. Use post shore jacks, screw jacks, or hydraulic jacks to lift the new beam into position beneath the joists. Finally, cut and install new columns to fit between the beam and the new footing.

2. Stiffening the Floor with Sister Joists Is a Tried-and-True Method

Adding a second joist of the same size alongside each existing joist, also known as sistering the joists, stiffens a floor. When headroom permits, sistering with taller joists provides more bang for the buck than sistering with same-size joists. Even though taller joists need to be notched to fit existing mudsills and support beams (see the left photo on the facing page), the added depth

Sister Joists

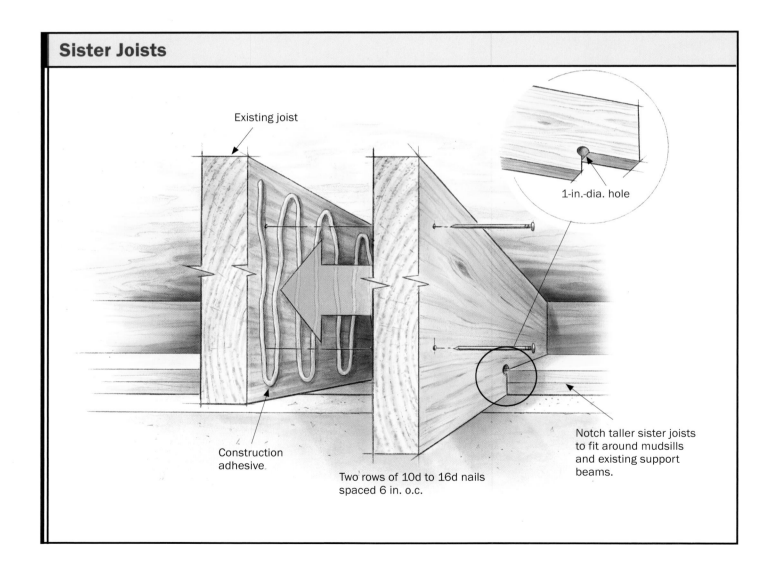

Existing joist

1-in.-dia. hole

Construction adhesive.

Two rows of 10d to 16d nails spaced 6 in. o.c.

Notch taller sister joists to fit around mudsills and existing support beams.

along the middle of the span provides extra support and further reduces bounce.

Engineered lumber—LVLs, for example—also can be used as sister joists and adds more stiffness to a floor than dimensional lumber.

If the existing floor joists are bowed downward noticeably, they might need to be jacked up slightly to make installing the new joists easier.

To minimize future squeaks, spread construction adhesive onto both the existing joist and the new joist. Position the top of the new joist alongside the top of the existing joist. Use a sledgehammer or a pry bar to force the bottom of the new joist along the mudsill and center support beam of the floor system (or the opposite mudsill on short spans) until it's flat against the existing joist. Nail the new joist to the existing joist with two rows of 10d to 16d nails spaced 6 in. o.c. For additional stiffness, sister joists can be applied to both sides of a joist.

Overcut notches are prone to splitting. Avoid runout by drilling a 1-in.-dia. pilot hole and then cutting up to it.

Make it flat. Use a sledgehammer to force the bottom of the new joist along the mudsill and center support beam of the floor system.

3. Flexible Plywood Strips Are Great for Tight Quarters

An alternative to adding full-length sister joists is to apply two layers of plywood to one side of the bouncy joists. Shorter, lighter, more flexible plywood strips often are easier to install in tight quarters than full-size dimensional lumber.

Rip ¾-in. plywood into 8-ft.-long strips equal to the height of the existing joists. Use a combination of construction adhesive and nails or screws to fasten two layers of strips to the existing joist, spanning from the mudsill to the center support beam or opposite mudsill. Place the first 8-ft.-long strip 1 ft. off center from midspan, and fill toward the ends with pieces. Then glue and fasten a second layer of plywood starting with a full strip 2 ft. off center from the first layer.

When fastening to joists less than 8 in. tall, drive two rows of 8d nails or 2-in. wood screws 6 in. o.c. into each layer of plywood. If the joists are taller than 8 in., add a third row of fasteners.

This system relies on good workmanship for success. Be generous with adhesive and fasteners, and make sure that the existing joists are solid and not deteriorated. If the connection between the two plywood layers and the existing joists isn't solid, you don't maximize the benefit of using this technique.

Plywood Strips

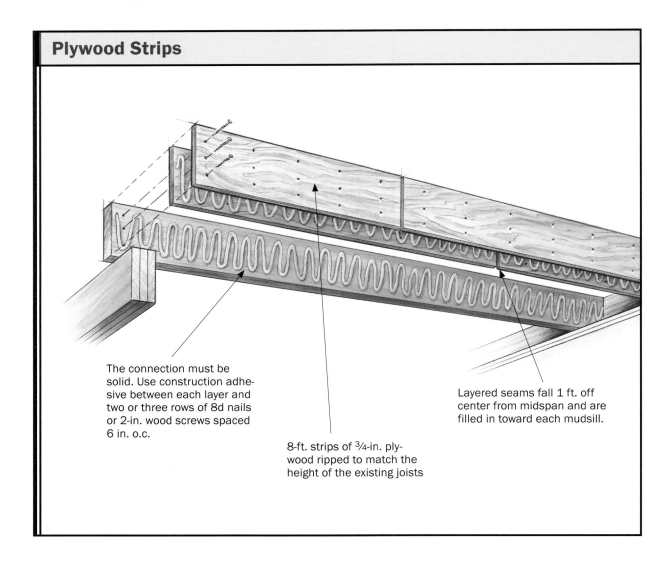

The connection must be solid. Use construction adhesive between each layer and two or three rows of 8d nails or 2-in. wood screws spaced 6 in. o.c.

8-ft. strips of ¾-in. plywood ripped to match the height of the existing joists

Layered seams fall 1 ft. off center from midspan and are filled in toward each mudsill.

4. Fix Weakened Planks and Second-Floor Bounces with a New Layer of Plywood

Floors in older homes often are decked with diagonally laid 1× planks instead of plywood or OSB. If you're planning to remodel or just add new flooring to a room, consider adding a layer of ¾-in. plywood subfloor sheathing over the lumber decking. When nailed through the old subfloor and into the joists, the new subfloor can help to reduce floor bounce. This solution also works for problematic second floors, where accessing the joists through the first-floor ceiling isn't a possibility.

When considering this option, think about the transition between old floor heights and new floor heights at doorways.

You may need to add thresholds and cut doors. Also consider the loss of headroom.

Sheathing also can be applied to the bottom of the floor joists instead of or in addition to being installed on top.

To minimize the chance of future squeaks and to help secure the new floor sheathing, start by spreading a generous amount of construction adhesive over the old planks. Lay the new sheets perpendicular to the floor joists, and orient panel ends over joists. Choose ring-shank nails or screws that are long enough to penetrate the joists by 1½ in., and space them every 6 in. o.c. Add an additional row of fasteners midway between the joists to pull the layers tightly together.

When installing on the underside of the joists, the process is the same. You can apply sheathing to the underside of floor joists only if the bottom edges are at the same level. If the joist level varies by more than ½ in., you can't use this method.

New Layer of Plywood

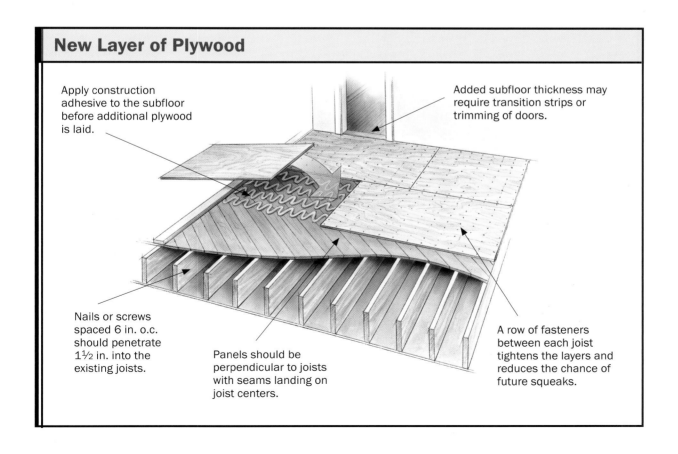

Apply construction adhesive to the subfloor before additional plywood is laid.

Added subfloor thickness may require transition strips or trimming of doors.

Nails or screws spaced 6 in. o.c. should penetrate 1½ in. into the existing joists.

Panels should be perpendicular to joists with seams landing on joist centers.

A row of fasteners between each joist tightens the layers and reduces the chance of future squeaks.

Perforated steel strap.

5. Metal Straps Are Deceptively Effective and Simple to Install

Tim Brigham of Koloa, Kauai, Hawaii, suggested the use of continuous steel straps to stiffen joists. Wrap the joists from the top of one end, around the bottom at midpoint, and back to the top of the opposite end. When a load is applied to the middle third of the joists, the steel strap transfers the force of the weight to the nails along the length of the joist, particularly along the ends where the joist is more rigid.

To begin, lift up the joists with jacks and a temporary beam about 1 ft. from midspan of the joists. Raising the joists slightly

(anywhere from ⅛ in. to ½ in., depending on the span and on the conditions) before installing this system helps to ensure that the straps are nice and tight when the jacks are removed.

Starting above the mudsill on one side of the joist, sink metal-connector nails through every hole along the first 2 ft. of strap. With the first 2 ft. fastened securely, the strap then should be fastened every 6 in. and folded with crisp bends around the bottom of the joist and onto the opposite side. Nail the strap to the opposite side of the joist the same way as the first side. For additional support, straps can be installed on both sides of the joists and cross-lapped at the center. Once all the joists are strapped, remove the temporary beam.

Metal Straps

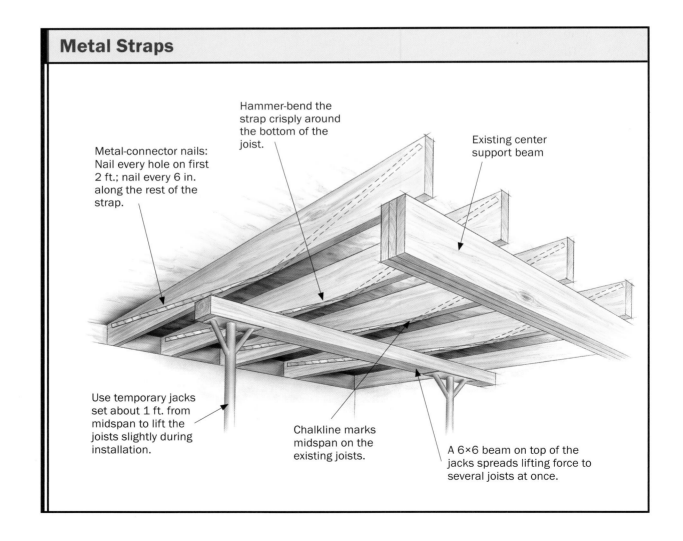

Metal-connector nails: Nail every hole on first 2 ft.; nail every 6 in. along the rest of the strap.

Hammer-bend the strap crisply around the bottom of the joist.

Existing center support beam

Use temporary jacks set about 1 ft. from midspan to lift the joists slightly during installation.

Chalkline marks midspan on the existing joists.

A 6×6 beam on top of the jacks spreads lifting force to several joists at once.

6. Solid Blocking Ties the Floor Together but Takes Time to Install Correctly

Properly installed solid-wood blocking helps to transfer weight to adjacent joists so that the floor acts as a stronger unified system. If you already have blocking or bridging installed between joists, it may be ineffective because it's not tight. Metal and wood cross-bridging are both prone to loosening over time as wood expands and contracts. Solid blocking is susceptible to shrinkage, but it typically works better than using the cross-bridging method.

This method works best if joists are dry, so it's a good idea to wait until late winter or early spring when the heating season is coming to an end and the moisture content of the joists likely will be at its lowest. Start by snapping a chalkline at the middle span of the floor running perpendicular to the joists. Using dry dimensional lumber, cut blocks just a whisker longer than the space between joists, and pound them into place so that they are tight. Blocks that aren't tight will end up causing squeaks.

Solid Blocking

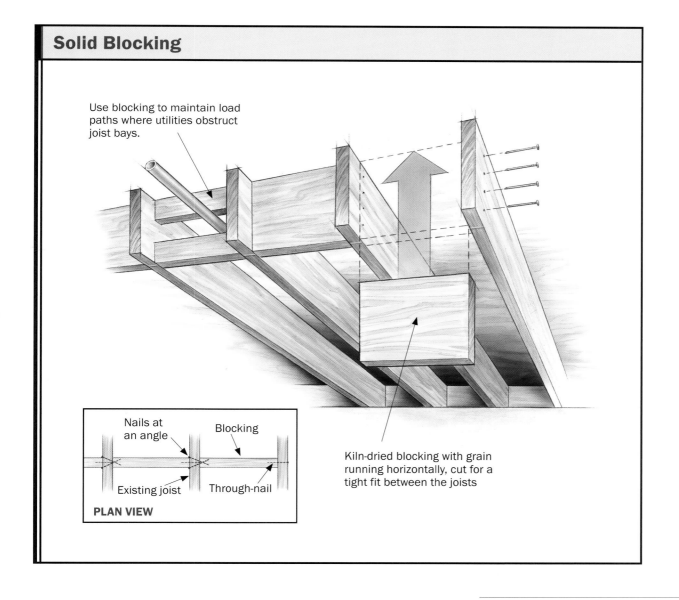

Use blocking to maintain load paths where utilities obstruct joist bays.

Kiln-dried blocking with grain running horizontally, cut for a tight fit between the joists

Nails at an angle

Blocking

Existing joist

Through-nail

PLAN VIEW

Sources

www.ablebuilders
.com
Long post-shore jacks

www.strongtie.com
Simpson Strong-Tie
CS20

It may be tempting to install this type of blocking in a staggered line because it's easier to fasten. But solid blocking is meant to work as a system, so keep the blocks in line. If you need to get around pipes or ductwork, use split blocks (a pair of 2×4s, for instance) on top and bottom to maintain the path.

Through-nail into the end of each block using three or four 16d spikes or 3½-in.-long wood screws. Pneumatic palm nailers work great for driving nails in these tight situations (see the photos at right and below). For added support, you also can install two or three rows of blocking spaced equally apart.

Prices are from 2007.

Mike Guertin *is a builder, remodeler, and contributing editor to* Fine Homebuilding *from East Greenwich, Rhode Island. His website is www. mikeguertin.com.* **David Grandpré** *is a registered professional engineer with C.A. Pretzer Associates, Inc., in Cranston, Rhode Island.*

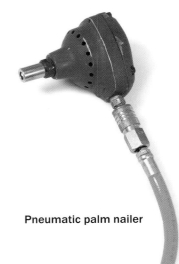

Pneumatic palm nailer

A pneumatic palm nailer is great for driving nails in spaces where swinging a hammer or fitting a full-size framing nailer isn't convenient.

Careful Layout for Perfect Walls

■ BY JOHN SPIER

Framing walls is one of the most fun parts of building a house. It's fast, safe, and easy, and at the end of the day it's satisfying to admire the progress you've made. Before cranking up your compressor and nail guns, though, you need to think through what you're going to do. You need to locate every wall precisely on the subfloor, along with every framing member in those walls (see the photo below).

Layout Starts in the Office

For one of our typical houses, layout and framing for interior and exterior walls start in the office a few days before my crew and I are ready to pick up the first 2×6. First, I review the plans carefully and make sure that all the necessary information is there.

Everything from rafters to kitchen cabinets fits better when you get the walls square and the studs in the right place.

I need the locations and dimensions of all the rough openings, not only for doors and windows but also for things such as fireplaces, medicine cabinets, built-ins, dumbwaiters, and the like. I also make sure the plans have the structural information I need for layout, such as shear-wall and bearing-wall details and column sizes.

At the site, one of Spier's many corollaries to Murphy's Law is that errors never cancel each other out; they always multiply. If the floor is anything but straight, level, flat, and square, the walls are going to go downhill (or uphill) from there. So before you get to layout, do whatever it takes to get a good floor, especially the first: Mud the sills, shim the rims, rip the joists. Sweep off the subflooring, and avoid the temptation to have a pile of material delivered onto it.

Take Layout Lines from the Mudsills

If the mudsills were installed perfectly square, you can avoid any discrepancies in the deck framing by plumbing up from the mudsills and measuring in from there.

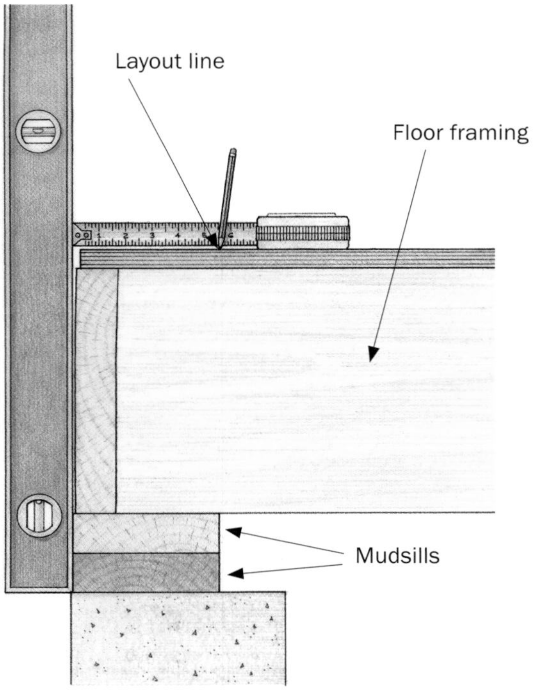

Layout line

Floor framing

Mudsills

Snap Chalklines for the Longest Exterior Walls First

I've learned over the years that it's best to snap the plate lines for the entire floor plan before building any of it. Problems you didn't catch on the prints often jump out when you start snapping lines.

When framing floors, I take great care to set the mudsills flush, square, and in their exact locations. Because the edges of floor framing and subflooring are not always perfect, though, I use a level to plumb up from the mudsills and establish the plate lines, measuring in the stock thickness from the level (see the illustration at left). I generally start with the longest exterior walls and the largest rectangle in the plan. When I have the ends of the longest wall located, I snap a line through the marks.

Once I've established the line for the first wall, I move to the parallel wall on the opposite side of the house. I measure across the floor from the first line to the opposite mudsill (again using a level to plumb up from the mudsill to the floor height) at both ends; if the lengths differ slightly, I use the larger measurement. I snap through these points, which gives me two parallel lines representing the long sides of the largest rectangle (see the illustrations on the facing page). It's okay if the plates overhang the floor framing by a bit, but I watch for areas that might need to be shimmed or padded—for instance, where a deck ledger needs to be attached to the house.

Establish the Right Angles

I locate three corners by measuring in from the mudsills. The fourth corner I locate by duplicating the measurement between the first and second because I need sides of equal lengths to create a rectangle. I check this rectangle for square by measuring both its

Setting the Stage for the Rest of the House

For the most precise wall layout, plot a series of rectangles that includes every wall. The larger the rectangle, the more accurate the wall position. Begin with the longest walls, and lay out the largest rectangle using diagonal measurements (see the photo at right). Working off established lines and square corners, work down to the smallest rectangle.

Equal diagonal measurements mean a square layout. After snapping chalklines for the longest parallel walls, the author takes corner-to-corner measurements to make sure the corners are square for a perfect rectangle.

1. Starting with the longest walls, measure and square the largest rectangle.

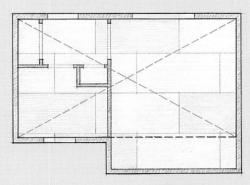

2. Working off those lines, plot the rectangle that includes the jog in the wall.

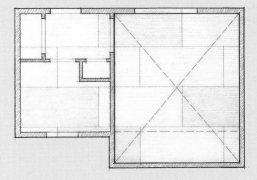

3. Now measure off the outside and form a rectangle for the longest interior wall.

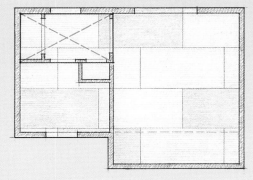

4. Last, form rectangles for the remaining interior walls.

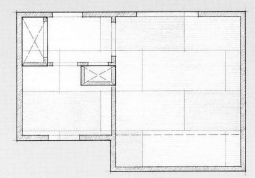

diagonals (see the photo on p. 175). If I've done everything right so far, the diagonal measurements should be very close, perhaps within ¼ in. I shift two corners slightly if I need to, making sure to keep the lengths of the sides exact until the diagonals are equal. Now, perpendicular lines are snapped through the corners, completing the rectangle.

Because I started arbitrarily with one long wall, I may find now that the rectangle, although perfectly square, is slightly askew from the foundation and floor. Also, some of the complicated foundations that I work on can have wings or jogs that are slightly off. If I can make everything fit better by rotating the rectangle slightly, I take the time to do it now.

Smaller Rectangles Complete the Wall Layout

With the largest part of the plan established, I lay out and snap whatever bays, wings, and jogs remain for the exterior walls. I use a series of overlapping and adjacent rectangles, which I can square by keeping them parallel to the lines of the original rectangle. I again check the right angles by measuring the diagonals.

Often, the plan calls for an angled component such as a bay. If these components are at 45 degrees, I lay them out from right angles by forming and diagonally bisecting a square. For angles that are not 45 degrees, I either can trust the architect's measurements on the plans, or I can use geometry and a calculator. The latter method is more likely to be accurate.

When all the exterior walls have been laid out, I turn my attention to the interior walls. Again, I start with the longest walls and work to the smallest, snapping lines parallel and square to the established lines of the exterior walls. I snap only one side of each plate, but I mark the floor with an X here and there to avoid confusion about where the walls will land. I also write notes on the floor to indicate doors, rooms, fixtures, bearing walls, and other critical information.

One last critical issue when reviewing the plans and laying out walls is watching for elements of the design that need to stay symmetrical. If the foundation contractor made one wing a bit wider than another, you don't want to build all three floors before realizing that the ridgelines of the two wings needed to match up. Make sure symmetrical elements are aligned at the first layout stage.

Chalkline tip. To snap chalklines for short walls, hook one end of the line to your boot and stretch the other end out to the mark. Rotating your foot slightly aligns the boot end, and you're ready to snap.

Make Plates from the Straightest Lumber

While I snap the walls, the crew is busy cutting and preparing material from the piles of stock. I have them set aside a pile of the straightest lumber. With the chalklines all snapped and with this material in hand, I start cutting and laying out the plates (top and bottom members) for the exterior walls (see the photo below). In this step of layout, we set the plates side by side on the layout line, and every wall-framing member is located and labeled. With this information, we assemble the walls on the floor, then raise them into place. I often call out measurements and have someone cut and hand up the material to keep mud, snow, and sawdust off the floor during this crucial phase.

As a rule, we plate the longest exterior walls to the corners of the house, and the shorter walls inside them. This approach sometimes needs to be modified—for instance, to accommodate structural columns, hold-down bolts, or openings adjacent to corners. Sometimes an obstruction or a previously raised wall dictates which wall can be built and raised first. The goal here is to build and raise as many walls as possible in their exact positions, especially the heavier ones. Moving walls after they're raised is extra work and no fun.

Before starting any framing, I established a common-stud layout for the entire structure based on two long perpendicular walls from which layout for the rest of the house framing can be measured (see the illustration on p. 178). This common layout keeps joists, studs, cripples, and rafters throughout the house vertically aligned from the foundation to the ridge, which makes for a strong, straight, and easily finished structure. We use this common layout to locate butt joints between pairs of plates because

Plates on deck. When wall positions are laid out, cut all the top and bottom plates (the long horizontal members) for the exterior walls, and place them on their layout lines.

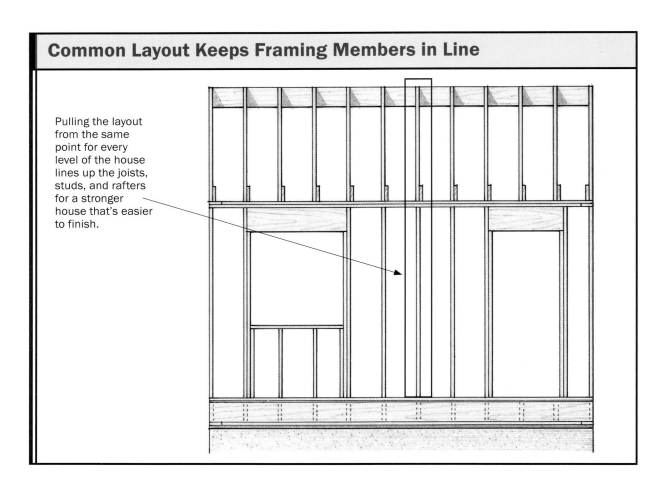

Common Layout Keeps Framing Members in Line

Pulling the layout from the same point for every level of the house lines up the joists, studs, and rafters for a stronger house that's easier to finish.

Copy the layout from the plate. To mark the cripple layout on the rough windowsill, just line it up on the plate and copy the layout.

code and common sense dictate that these joints land on a stud or a header.

As my crew and I measure and cut the pairs of wall plates, we lay them on edge along their layout lines, sometimes tacking them together with just a few 8d nails to keep the plates held together and in place.

Window, Door, and Stud Layout at Last

When all the exterior plates are in place, it's finally time to lay out the actual framing members. I always start with the rough openings for windows and doors. Most plans specify these openings as being a measured distance from the building corner to the center of the opening, which works fine. You can allow for the sheathing thickness or not, but once you choose, be consistent,

especially if openings such as windows have to align vertically from floor to floor. Obviously, if an opening such as a bay window or a front door is to be centered on a wall, center it using the actual dimensions of the building, which may differ slightly from the plan.

Rough openings are a subject worthy of their own chapter, but in a nutshell, I measure half the width of the opening in both directions from the center mark. I then use a triangular square to mark the locations of the edges of trimmers and king studs, still working from the inside of the opening out. Various other marks, such as Xs or Ts, identify the specific members and their positions (see the sidebar on p. 180).

Next, I mark where any interior-wall partitions intersect the exterior wall. At this point, I just mark and label the locations; I decide how to frame for them later. I also

Stud layout is always taken from the same two walls. One crew member holds the tape at wall offset while the other marks the stud position (photo above). Even when there is a break in the wall, the layout is pulled from the same place to keep all the framing aligned (photo at left).

Laying out multiples. For things such as short closet wall plates, line them up and draw two walls' worth of layout lines at once.

Framer's Shorthand: What Those Little Marks Mean

When the framing members are marked, a full-length stud is indicated by an X. A trimmer or jack stud is a T or J, and a C or X indicates a cripple (a short framing member below a sill or above a nonstructural header). Other framing, such as partition posts and corner posts, are labeled, along with any special framing instructions.

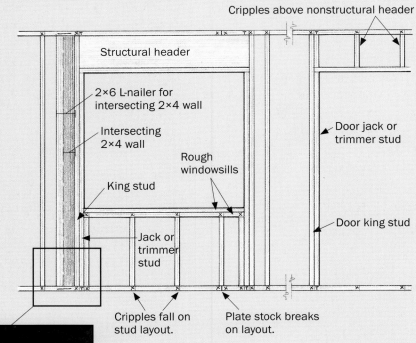

Cripples above nonstructural header

Structural header

2×6 L-nailer for intersecting 2×4 wall

Intersecting 2×4 wall

Rough windowsills

King stud

Jack or trimmer stud

Door jack or trimmer stud

Door king stud

Cripples fall on stud layout.

Plate stock breaks on layout.

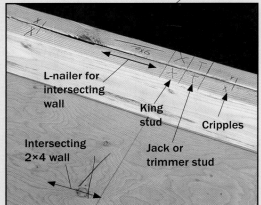

L-nailer for intersecting wall

King stud

Cripples

Intersecting 2×4 wall

Jack or trimmer stud

Spell out what you need on the wood. Mark key areas where studs might need to be left out to allow installation of things too wide to be carried though the doors, such as one-piece tub/shower units.

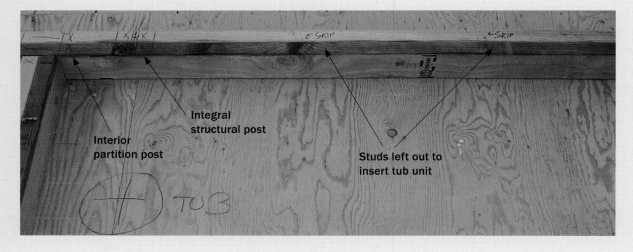

Interior partition post

Integral structural post

Studs left out to insert tub unit

locate and mark any columns, posts, or nailers that need to go in the wall. I lay out any studs that have to go in specific locations for shelf cleats, brackets, medicine cabinets, shower valves, cabinetry, ductwork, and anything else I can think of. Doing this layout now is much easier than adding or moving studs later.

Finally, I lay out the common studs on the plates. Studs are commonly spaced either 16 in. or 24 in. o.c. to accommodate standard building products. By doing the common-stud layout last, I often can save lumber by using a common stud as part of a partition nailer. I almost never skip a stud because it's close to another framing member, which, I've learned the hard way, almost always causes more work than it saves. I occasionally shift stud or nailer locations to eliminate small gaps and unnecessary pieces. I keep the plywood layout in mind here, though, so that I can use full sheets of sheathing as much as possible.

Inside Walls Go More Quickly

Once the exterior walls are built and standing, I cut the interior-wall plates and set them in place. Where two walls meet, I decide which one will run long to form the corner so that the walls can be built and raised without being moved. Also, facing a corner in a particular direction often provides better backing for interior finishes, such as handrails or cabinetry, and sometimes is necessary to accommodate such things as doorways or multiple-gang switches.

When the plates are cut and set in place, I do the stud layout. Just as with the exterior walls, I do the openings first, then nailers and specific stud and column locations. Next, I mark the locations of intersecting walls and finally overlay the common-stud layout on the plates.

Where Walls Come Together

Where one wall meets the middle of another, I use a partition post if the situation dictates it, but more often, I opt for an L-nailer. To make an L-nailer, I use a wider stud on the flat next to a common stud whenever possible. It's faster and easier; it accommodates more insulation; and it saves the subs from drilling through those extra studs and nails. If I use U-shaped partition posts (a stud or blocks on the flat flanked by two other studs) in an exterior wall, I need to make sure to fill the void created by the partition post with insulation before the sheathing goes on.

With the interior plates all there, we can nail in the studs, raising walls as we go. I mark key areas where studs should be crowned or specially selected, such as areas with long runs of cabinetry, and also studs that might need to be left out to allow installation of things too wide to be carried though the doors. I also nail double top plates to as many walls as possible if they don't interfere with the lifting process.

John Spier is a builder on Block Island, Rhode Island. His book, Building with Engineered Lumber, *is available from The Taunton Press, Inc.*

By doing the common-stud layout last, I often can save lumber by using a common stud as part of a partition nailer. I almost never skip a stud because it's close to another framing member, which, I've learned the hard way, almost always causes more work than it saves.

Setting the Stage for Wall Framing

■ BY JIM ANDERSON

The cutlist for a house that I frame today could have 50 different header sizes (not to mention multiples of the same size) with 200 cripples in 15 different lengths. To do good work on increasingly complex projects in a reasonable amount of time, I've had to get organized. Regardless of size, though, every project has its own web of problems, and being better organized can only make my work flow more smoothly.

I've also found that the job site is neater as I become better organized. I once landed a framing job because the customer liked the clean job site. Being smart about keeping debris and stacks of lumber off the floor yields fewer accidents and reduces a lot of the bending that makes my joints ache.

Double-Check the Floor Plan to Avoid Making Any Mistakes

In a world where all floor plans were perfect, I could, without checking the floor plan to verify its accuracy, simply transfer all of its

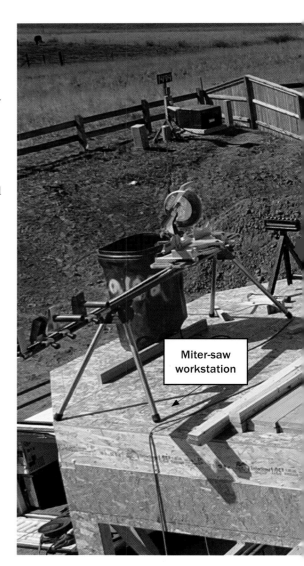

Miter-saw workstation

Rule Out Floor-Plan Errors

1. Compare the floor plan to the structural plan and the window schedule. Header types and sizes are listed on the structural plan, and window types and rough openings are listed on the window schedule. Make sure everything matches.

2. Mark up the floor plan with details about each door and window opening.
3. Transfer the framing details to the cutlist.
4. Use your well-organized cutlist to group wall components on the floor for easy identification.

Bearing headers

Long cripples

Nonbearing headers

Medicine cabinets

Sills

3:10 7487 VIKING

measurements to the cutlist. Just because the drawings look right, however, doesn't mean that they are.

I double-check my floor plan against two other documents, the structural plan and the window schedule. The structural plan spells out the type and depth of headers as well as the size of the posts that support them (for example, two 2×4s). Checking the floor plan against the window schedule and the structural plan roots out any hidden problems. Once all the details about framing members have been checked and transferred to the floor plan, I use the new marked-up version to create a cutlist. The list includes dimensions for every piece of wall framing that needs to be cut except the trimmers, or jack studs. I scribe and cut those last to account for any discrepancies in header widths.

Organize the Cutlist with the Largest Framing Members First

I begin the cutlist with headers and arrange them in descending order from widest to narrowest. I cut the list in the same order for two reasons: It gets the biggest pieces out of the way first, and in case someone on the crew is ready to move on to framing the outside walls before the cutlist is finished, the big headers will be cut and ready to go in the outside walls, where many of them are needed.

Starting at one corner of the plan, I circle around until I've accounted for all the headers. I double-check by counting the number of headers on my list, then count the openings that require headers. If the numbers match, I'm done, and I go on to check the sills using the same method.

After listing the headers and sills, I list the lengths for special studs (other than precut wall studs), interior headers, all cripples, and medicine-cabinet rough openings, which we preassemble.

Everything is listed in order of descending length. Cutting the longest stock first means that if I make a mistake, it can be chopped into shorter stock for use later in the list. (I still haven't saved enough money for one of those board stretchers.) Then I type the list into the computer and print it out so that no one has to waste time deciphering my not-so-perfect penmanship.

Cut and Assemble Headers at a Comfortable Height

Because there's a lot of bending and lifting of heavy material in the course of a day, I set up my miter saw and cut as much material as possible at a comfortable working height.

Building materials usually arrive in bundles. If the LVLs and wide 2× material are on top of a bundle of plywood, I have the bundle delivered fairly close to where I'm setting up. If the LVLs arrive bundled separately, I move them onto a pair of sawhorses nearby. Either way, they're off the ground at a good height for cutting and assembling.

I measure and mark all headers to length before cutting. With common dimensional lumber, usually Douglas fir or hem-fir, I stack all the lumber with crowns facing away. With LVLs, I square one end as I stack them. Then I go down the cutlist, marking measurements and the crown direction with a large arc on the common lumber. The longer I stick with one task, measuring everything, then marking and cutting, the faster I get it done.

I mark the longest headers first, but if I can get more than one header out of a piece

Organize your work in assembly-line fashion. Whenever possible, measure and mark, then cut and preassemble more than one piece at a time.

Don't mark one board at a time. The author uses a framing square to strike lines for cripples across the tops of eight headers at once. Lying within easy reach, the cripples will be nailed to the headers at a comfortable height atop the plywood; then the headers will be stacked on the floor nearby.

of material, I mark shorter headers at the same time. After marking everything, I cut the headers to length and stack them for assembly.

For many houses, I need to join a pair of 1¾-in.-thick LVLs to make a single long header. I glue together each pair with three beads of construction adhesive, then hammer a couple of nails every 4 ft. or so to draw them tightly together before finishing the assembly with a nail gun.

The nails and construction adhesive don't necessarily add strength, but they improve adhesion between the two pieces. One rainy spring, some LVL headers cupped badly when their outer portions absorbed water faster than the interiors. Construction adhesive is cheap insurance against unnecessary repair.

Pairs of dimensional headers are nailed together similarly except they have ½-in. OSB spacers between them so that they'll match the thickness of a 2×4 wall. I rip several 3-in.-wide strips for spacers using the base of my circular saw to gauge the width. Then I chop a bunch of them on the miter saw to lengths ½ in. less than the width of their corresponding headers (8¾ in. for a double 2×10 header). Before nailing headers together, I place OSB spacers every 16 in. These 3-in.-wide spacers can be used with headers as small as 3½ in. wide. For the 3½-in. headers, I run the spacers lengthwise.

I clearly mark each header's measurement on one face of the header and stack them side by side (with measurements facing up) in order of descending width and length. This process allows any of us on the crew to read and grab what we need quickly. There's no need to use a tape measure or to dig through a pile for the header that is always on the bottom.

Cut and Label 2×4 Headers, Sills, and Cripples

Now it's time to cut the little pieces: 2×4 stock including sills, cripples, nonbearing headers, medicine-cabinet frames, and special-length studs. With multiple pieces of the same size to cut, rather than using a square and pencil and cutting with a circular saw, I set the adjustable stops to the correct length and make repetitive cuts on my miter saw (see the sidebar on p. 188). I position the saw stand close to the lumber pile and cut the longest pieces first—special studs and long sills—from the straightest stock available. I save the more twisted or crooked stock for shorter sills and cripples.

Special studs and sills are marked for length and stacked neatly in descending order. Sills get a line beneath the measurement. Longer cripples are marked for length, and then they're stacked adjacent and perpendicular to the bearing headers.

After cutting nonbearing headers and shorter cripples to length, I mark them and stack them on sawhorses or on a pile of nearby sheathing to be assembled.

A router removes exterior sheathing at window and door openings. To follow the framing around the opening, you need a plunge-cutting straight bit with a bearing beneath the cutter.

Double Your Speed with a Miter Saw

For 14 years, I used a tape measure and a square for marking all the pieces on a cut-list—up to 200 pieces on complicated framing jobs. I worked as efficiently as possible, but it still took a long time to get everything cut and stacked.

Eight years ago, I started using a miter saw, and its ability to cut two pieces at once doubled my speed the first day. Now, I have a miter-saw stand with adjustable stops. I set the stop to the length that I want and start feeding lumber, often cutting three pieces at a time (see the photo below left).

I've also scribed often-used lengths, too short for the adjustable stops, onto the sur-face of the saw table or highlighted the fence with a red marker. Now the only mark that I make on the lumber is the length and a note indicating whether it's a sill, a cripple, or a header.

I've heard other framers complain that it takes too long to set up the miter saw and stand each day. I'd argue that it takes only a little longer than pulling out the circular saw and cord, and that the time is made up easily from the increased speed, accuracy, and capacity of the saw—even more so when it comes to repetitive cuts.

Leave your square and tape measure in your tool belt. Once the adjustable stops are secured the proper distance from the blade, cutting multiples with the miter saw is fast and accurate.

Add Cripples to Nonbearing Headers When You Build Them

Because everything is now cut to length and within easy reach (except trimmer studs), it makes sense to assemble the interior headers. All the double 2×4 headers are assembled at this time with an OSB spacer laid lengthwise between them. As the headers are built, I mark locations for cripples by grouping several of them together, and using a framing square, I mark 16-in.-o.c. layout lines along their tops. Referring back to the cutlist, I locate the proper cripples and then toenail them to the corresponding header.

I assemble all the nonbearing headers at this time for a couple of reasons. I think that it's easier to nail the cripples to the headers on a waist-high table rather than bent over or up on a ladder. And when I'm stocking the wall with parts rather than grabbing a header and then looking for a matching handful of cripples, I can lay the whole assembly in place. For the nonbearing headers, I don't worry about aligning their cripples with the 16-in.-o.c. wall layout. After building and attaching the cripples, I mark the length of each header on the bottom and stack them in descending order (length facing up) next to the long cripples that are on the side opposite the bearing headers.

Nonbearing headers that are shorter than 35 in. are made of a 2×4 laid flat. They have three cripples, each nailed through the header: one cripple on each end and one in the center. I mark the length on each and stack them next to the other nonbearing headers with the length visible, facing up.

I also assemble medicine-cabinet frames. They consist of two 16-in. blocks with a 24-in. block between them, nailed together in a U-shape. I label the frames MC and stack them next to the nonbearing headers with the label facing up.

With Studs and Headers Assembled, Building the Wall Is a Snap

While one of my helpers finishes the cutlist, I finish installing the wall plates and work my way around, marking them up with information from the floor plan. Then we begin stocking the first wall with studs and components from the stack. If we come to a window opening that reads "54 2×10 5/20" on the plate, that means the opening calls for a 54-in. 2×10 with 5-in. cripples above the header and 20-in. cripples below the sill.

For the exterior walls, all that's left is to cut the trimmer studs. I nail the headers to the king studs and scribe the trimmers to fit, which accounts for any inconsistency in the header stock. Because I stop only to cut the trimmers at the end, the wall goes together quickly, and there's less cutoff debris lying around.

As I nail the wall together, I start at one end, with a corner if there is one, and continue nailing every stud or cripple along the bottom of the wall. Then I nail the studs and cripples to the top plate in the same way, starting at one end and working my way to the other end. Finally, I add the second plate to the top plate. Nailing in this way is orderly, so it's easier for me to keep track of what has been nailed, making it less likely that I'll forget to nail something.

Jim Anderson is a framing contractor in Littleton, Colorado.

TIP

By stopping only to cut the trimmer studs at the end, the wall goes together quickly, and there's less cutoff debris lying around on the floor.

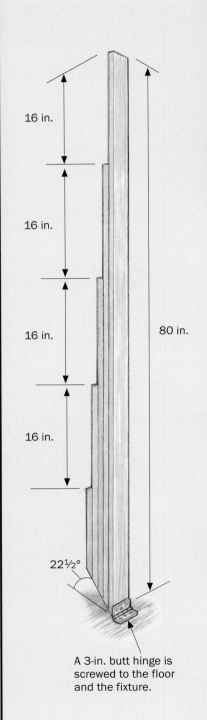

16 in.

16 in.

16 in.

16 in.

80 in.

22½°

A 3-in. butt hinge is screwed to the floor and the fixture.

Sheathe the wall as it lies flat.

Sheathed walls are heavy. A simple wall-raising fixture made of overlapping 1×4s supports the rising wall, allowing hands to get a better grip.

Laying Out and Detailing Wall Plates

■ BY LARRY HAUN

As a beginning carpenter, I was so afraid of making a mistake when laying out walls that the process took me hours. Layout seemed like an exact science, one whose rules I didn't know.

Transferring measurements from blueprints to full-size markings on the floor and then cutting the wall plates to fit these markings sets the stage for all of the carpentry that follows (see the illustration on the facing page). If the walls are out of parallel or if rooms are not square, roof rafters, cabinets, and even floor tiles won't fit properly.

Done correctly, though, wall layout leaves a set of clearly marked plates—templates, in a sense, that can enable relatively inexperienced carpenters to assemble the walls. Well-laid-out walls compensate for errors in the floor, be it slab or frame, and ensure that fixtures such as bathtubs fit between the walls. Once I learned the rules, learned where being off ¼ in. matters and where it doesn't, I relaxed and then began to enjoy laying out walls.

The Plans May Not Tell the Whole Story

Before beginning to frame a house, I study the plans at home. I note what the plans show and what must be inferred. There are often differences between plan dimensions and what is needed on the job. For example, plans may call for 36 in. between walls for a set of stairs (see the top illustration on p. 194). So that the stairs themselves will measure a full 36 in., I leave at least 38½ in. from stud wall to stud wall. The extra 2½ in. leaves enough room for ½-in. drywall and a ¾-in. skirtboard on each wall, enclosing 36-in. treads and risers between them.

To accommodate a bathtub, most bathrooms need to be 60 in. wide. I usually frame bathrooms ⅛ in. or ¼ in. big so that the tub can be installed with ease. Nonstandard-size tubs are becoming more common, and in these cases, the tub supplier may be a more reliable source for the tub's dimensions than the plans.

King and Jacks Are More Than a Poker Hand

They hold up your house as well. Through walls are continuous parallel walls that run the house's long dimension. Butt walls fit between them. Cripples, studs, headers, plates, and channels comprise both types of wall. The drawing below is a glossary of these builder's terms.

Double top plate goes on after the wall is framed; it laps and ties intersecting walls together.

Headers bridge openings to carry loads from above.

King stud ties door or window framing together.

Trimmer or jack stud supports header.

Top and bottom plates establish wall length.

Cripples are short studs that fill in above headers and below windows.

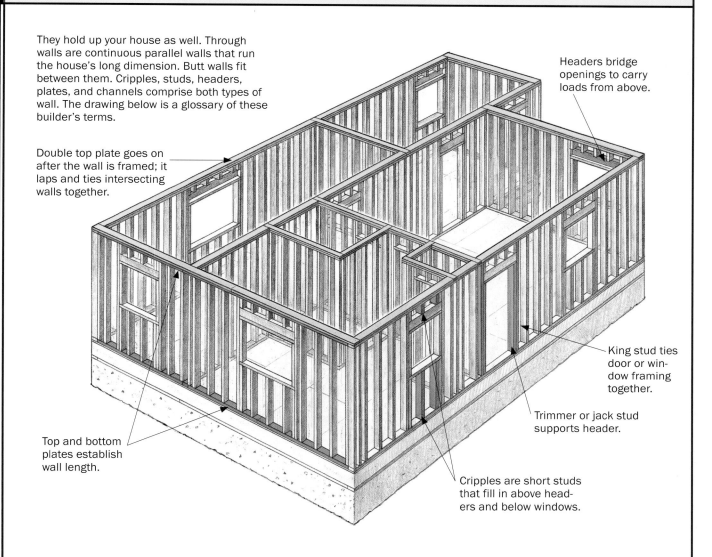

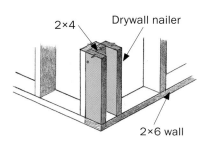

TRADITIONAL THREE-STUD CORNER

2×4

Drywall nailer

2×6 wall

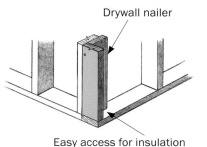

TWO-STUD CORNER

Drywall nailer

Easy access for insulation

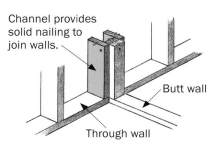

THREE-STUD CHANNEL

Channel provides solid nailing to join walls.

Butt wall

Through wall

Plans May Call for a 36-in. Stairway, but That Often Means 38½ in. between the Studs

36-in. tread

¾ in. skirtboard

½ in. drywall

38½ in.

Plans often call for halls to have 36 in. between finished walls. To achieve this dimension, the studs and plates must be 37 in. apart, allowing for ½-in. drywall on each wall. For that matter, 36-in. halls, common on older stock plans, don't allow adequate room for code-required 2/8 (32-in.) doors and their trim.

I lay out halls ending at a 2/8 door with 40 in. between stud walls, which leaves room for a 37-in. header with a king stud on each side (see the illustration below). Typically, the extra inches are stolen from bedrooms on both sides of the walls, a theft that is hard to notice.

I check that all the plan dimensions are correct. It's not uncommon for the room dimensions and the thicknesses of their walls to add up to a different number than that called out as the overall size of the

Code Specified 2/8 Doors Don't Fit Well in 36-in. Hallways

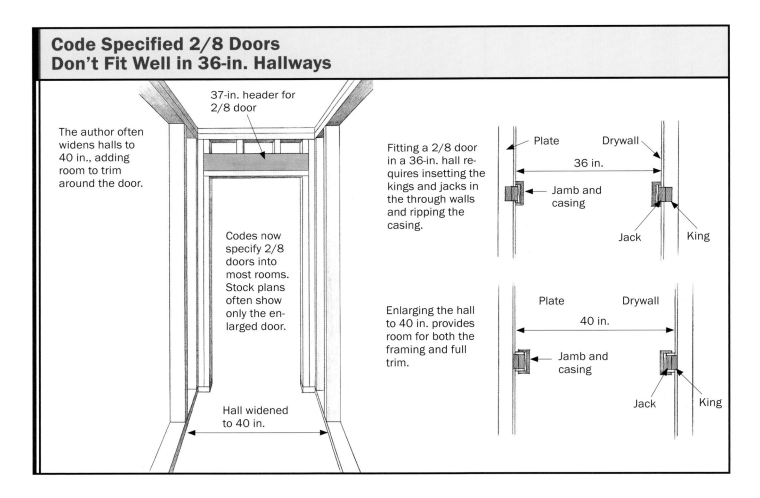

37-in. header for 2/8 door

The author often widens halls to 40 in., adding room to trim around the door.

Codes now specify 2/8 doors into most rooms. Stock plans often show only the enlarged door.

Hall widened to 40 in.

Fitting a 2/8 door in a 36-in. hall requires insetting the kings and jacks in the through walls and ripping the casing.

Plate

Drywall

36 in.

Jamb and casing

Jack

King

Enlarging the hall to 40 in. provides room for both the framing and full trim.

Plate

Drywall

40 in.

Jamb and casing

Jack

King

house. Figuring out these discrepancies is much easier at my kitchen table than in the field. I note which walls are 2×4 (most interior walls) or 2×6 (most exterior walls and walls with plumbing). I make notes directly on the plans so that all the needed information is in one place, and I discuss any wall-location changes with the owner, builder, or architect before snapping the wall lines.

Although following plan dimensions is important, most builders view plans as a guide and not as law written in stone. For example, when working on a concrete slab, you might find that a pipe will be 1 in. or so off layout (see the illustration below left). Fixing this error would be a huge job for the plumber, and it would slow the builder's schedule. Rather than snap the wall line to the plan dimension, you can usually move the wall to cover the pipe. Check first, though, that moving this wall doesn't affect adjoining areas where space is crucial.

Check Floors for Parallel and Square

Once you're familiar with the plans, it's time to sweep the floor and to begin laying out walls. The first step always is to check that the floor is parallel and square (see the illustration below right). I measure the floor from outside to outside at both ends of the longest exterior walls. If the measurements are equal, I mark the inside of the walls on the floor using a scrap of 2×4 or 2×6 plate stock as my guide. Pencil shows up well on wood, but I use keel, or lumber crayon, on concrete.

From these marks, I snap chalklines that establish the insides of the long walls (see the illustration on p. 196). An awl driven into a wood floor holds the end of a tape or chalkline. I use a weight on concrete slabs; in green concrete, you can drive a nail into the slab.

see the illustration on p. 196

Adjust Wall to Fit Plumbing

Sometimes the plumber gets it wrong. Moving walls a few inches often puts pipes in the right spot.

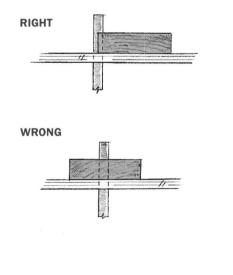

RIGHT

WRONG

Squaring Plates on an Out-of-Parallel Slab

Check first that the two longest sides of the deck or slab are parallel. If they aren't, you'll need to adjust them, and snap the chalklines, marking these plates parallel.

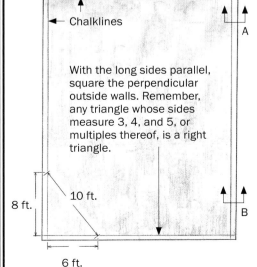

← Chalklines

A

With the long sides parallel, square the perpendicular outside walls. Remember, any triangle whose sides measure 3, 4, and 5, or multiples thereof, is a right triangle.

B

8 ft.

10 ft.

6 ft.

SECTION A
Plates can be inset by the thickness of the sheathing, or more with some sidings, to square up walls.

SECTION B
Plates can hang over the deck as much as 1 in.

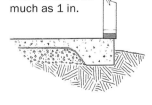

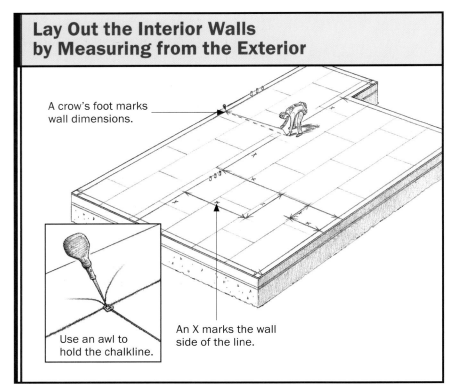

A crow's foot marks
wall dimensions.

Use an awl to
hold the chalkline.

An X marks the wall
side of the line.

scrap as a guide. After connecting these marks with a chalkline, I check the corner for square by measuring 3 ft. along one wall and 4 ft. along the other; then I check this triangle's hypotenuse. It will measure 5 ft. if the corner is square. Multiples of 3, 4, and 5 work as well, so a 6-8-10 triangle is also a right triangle.

If a corner is slightly out of square, you can adjust it much as with parallel walls. Move one end of the wall out a bit and the other end in. Take special care to make sure the kitchen and bathrooms are square so that cabinets and tile can be set easily.

Measure from the Exterior Walls to Lay Out the Interior Walls

Measuring from the long exterior walls, I can now mark the parallel walls on the floor. A plan dimension might read 12 ft. 6 in., for example, outside to center of a 2×4 wall. To locate the edge of a wall, simply move the mark over 1¾ in., or half the thickness of a 2×4. I locate both ends of an interior wall by measuring from an exterior wall and marking a V or crow's foot on the floor. Long walls such as those for bedrooms and hallways are laid out before the short walls for closets and bathrooms are located.

I always make an X with keel alongside the crow's foot to show on which side of the mark the plate will fall. Be careful: An X on the wrong side of a bathroom wall, for example, may mean that the tub won't fit. Some carpenters snap two chalklines to note that the plate will fall between the lines. This seems to be an unnecessary extra step.

All of the chalklines should be snapped straight through wall openings. Pay no attention to door and window openings when snapping chalklines.

I note anything out of the ordinary on the floor with keel. For example, I indicate the end of a short wall with a mark and write "end" at that point. If a plumbing wall is to be 2×6, I write that on the floor. I try to

Tweaking Walls Square on an Out-of-Square Deck

If the exterior walls are slightly out of parallel, adjustments can be made. The last house I helped to frame was built on a slab whose long sides were 1 in. out of parallel. I moved both walls inward ¼ in. at the wide end of the slab, decreasing the width ½ in. At the narrow end, I moved each wall out ¼ in., increasing the width ½ in. and making the walls parallel. Sometimes nothing you can do will get the walls perfectly parallel. If I can get them to within ¼ in. over 12 ft., that's usually good enough.

Keep in mind what will cover the exterior walls. Wall coverings often extend below the framing onto the slab or foundation by 1 in. or so. If you move a wall in on the foundation too far, the siding may not be able to extend below the sill.

Snapping lines for the perpendicular exterior walls comes next. I mark the wall location on the floor at each end of the first perpendicular wall, again using plate

Plate the Longest Walls through from End to End

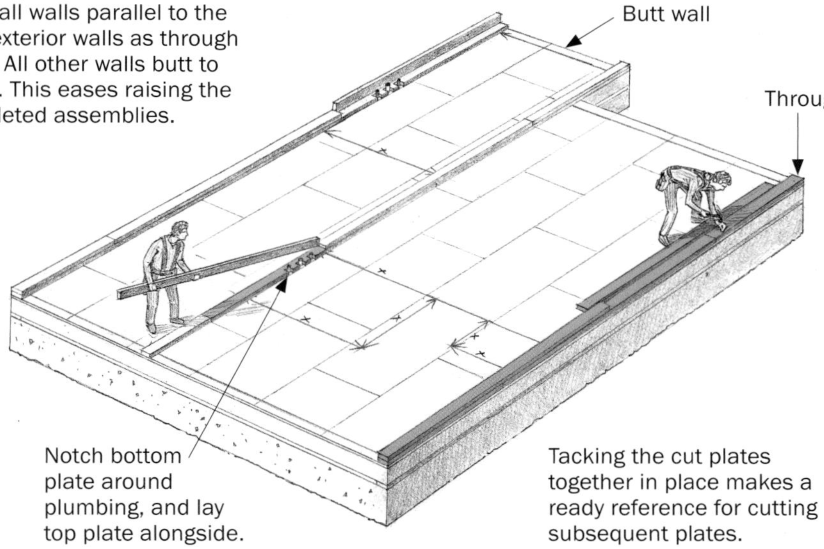

Plate all walls parallel to the long exterior walls as through walls. All other walls butt to these. This eases raising the completed assemblies.

Butt wall

Through wall

Notch bottom plate around plumbing, and lay top plate alongside.

Tacking the cut plates together in place makes a ready reference for cutting subsequent plates.

TIP

If you snap a line in error, make a clear correction. Rub out an erroneous chalkline with your foot or draw a wavy line through it before snapping another. To avoid confusion, you can snap a second chalkline using a different color chalk.

keep it simple because too many marks can be just as confusing as too few.

Distribute All the Plate Stock before Cutting Any to Length

After I finish snapping all the lines, the next step is carrying and placing two pieces of plate stock along every wall line (see the illustration above). These pieces of stock will be the top and bottom plates, and their lengths should approximate the wall length. To ensure straight and strong walls, I use long (16 ft., if possible), straight stock for plates. It's a good idea to sight the plate stock and to use the straighter pieces for the top plates. The bottom plates can be easily straightened when they're nailed to the floor, but the top plates must be braced straight after the walls are raised.

Once the plate stock is distributed, I start plating the longest outside walls. These plates run through from corner to corner and are called through walls. Walls that fit

between other walls are called butt walls. The chalklines mark the exterior walls' inside edge. Because the chalklines represent the inside of the walls, I locate the exact ends of the first through walls by lining up the end of the plates on a scrap of plate stock held on the intersecting wall's chalkline and cut the plate to length.

I tack the bottom plate to the floor with two or three 8d nails, stack on the top plate and tack it to the bottom plate. I continue stacking and tacking until I reach the far end of the wall. There, I mark the ends of the plates as I began, with a scrap held to the butt wall's chalkline, and cut the plates to length.

Staggering joints in the top and bottom plates is okay, but not necessary. The bottom plate will be nailed to the floor. After the studs are nailed in, a second, or double, plate will be nailed to the top plate, reinforcing it and tying together the walls. At least 4 ft. must separate joints in these top plates. So that intersecting walls can attach solidly, I keep joints in both top plates 4 ft. from intersections.

Quickly cut through walls. Using a circular saw, line up cuts on butt-wall plates with the intersecting through walls.

Anchor-bolt marker. Lay the plate exactly on the opposite side of its chalkline, center the bolt in the marker, and then whack it with a hammer to mark where to drill the plate. The second bolt is for 2×6 plates.

Once the exterior walls are done, I cut, stack, and tack (with 8d nails) the plates for the interior walls that parallel the long exterior walls. All perpendicular walls will butt into these through walls. I run all plates continuously, ignoring door and window openings. I cut the bottom plate from door openings later.

Cut the Plates to Length by Eye

The wall lines snapped on the floor show the position and length of each plate. It takes a bit of courage and practice, but you should be able to cut wall plates square and to length without further use of measuring tape or square. For through walls, sight the circular-saw blade on the chalkline below. Line up cuts on butt-wall plates with the intersecting through walls (see the photo above). If you have a 10-in. saw, you can tack the plates together and cut them with one pass.

The bottom plates can be cut a bit short. The result will be a harmless gap where two bottom plates intersect. The top plates, though, must be cut within 1/16 in. Otherwise they'll push or pull walls out of plumb.

Plating on a concrete slab is much the same as on a wooden deck, but there are a few differences. Remember to use treated wood for the bottom plate. The bottom plate can't be as easily tacked to a slab as it can be to a wood deck. I simply lay the bottom plates on their lines and tack on the top plates. I toenail intersecting plates together so that they don't move until I'm ready to nail together the walls.

Often, the bottom plate of an exterior wall must be bolted fast to a concrete slab. Typically, anchor bolts are cast into the edge of the slab, and holes are drilled in the plate to receive the bolts. I mark the holes with a bolt marker (see the bottom photo on the facing page). Set the plate directly on the chalkline, but on the opposite side of where it will finally go, and mark the hole location. Interior walls on slabs are usually nailed home with powder-driven pins or with concrete nails. Codes are specific about what fasteners can be used, so if in doubt, ask your building department.

When plumbing is roughed in, as it is on a slab, plates can't be stacked. I notch the bottom plate to fit around the pipes, lay it on the line and put the top plate next to the bottom.

Well-Marked Plates Mean Carpenters Don't Need to Think

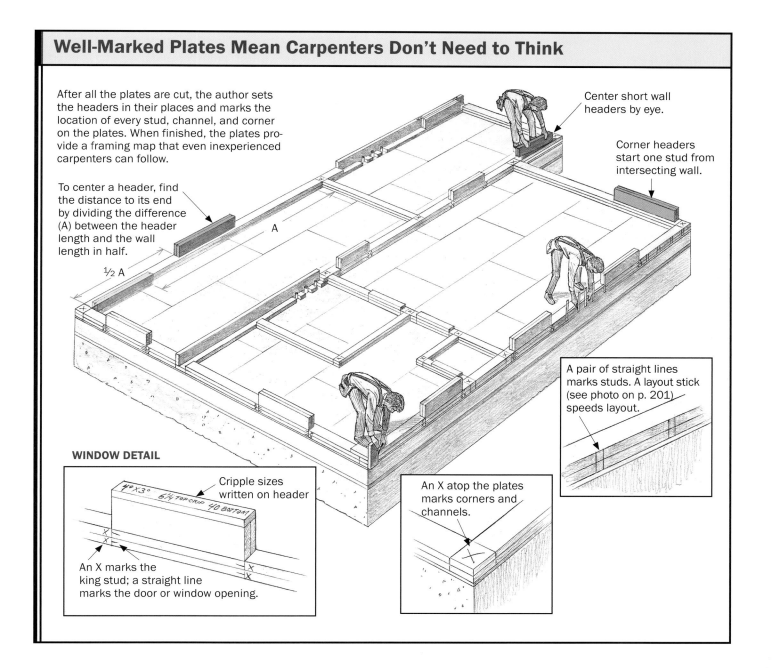

After all the plates are cut, the author sets the headers in their places and marks the location of every stud, channel, and corner on the plates. When finished, the plates provide a framing map that even inexperienced carpenters can follow.

To center a header, find the distance to its end by dividing the difference (A) between the header length and the wall length in half.

½ A

A

Center short wall headers by eye.

Corner headers start one stud from intersecting wall.

A pair of straight lines marks studs. A layout stick (see photo on p. 201) speeds layout.

WINDOW DETAIL

Cripple sizes written on header

An X marks the king stud; a straight line marks the door or window opening.

An X atop the plates marks corners and channels.

Channel marker. This light-weight 3½-in.-wide tool enables carpenters to mark intersecting butt walls on all three sides of through walls with a pencil. A similar tool can be quickly site-built of plate scrap.

Mark Every Stud on the Plates

Detailing plates is marking the location of all corners and intersecting walls, headers, and studs in every wall on the plates (see the illustration on p. 199). Because hundreds of marks are made on the plates, I keep this marking system simple. Extra marks are confusing.

I start detailing by marking the locations of corners and channels (three-stud assemblies in through walls to which butt walls are nailed). These spots require extra studs so that walls can be nailed together properly. The extra studs also provide backing for drywall inside and sheathing outside.

I lay out corners and channels by lining up a channel marker (see the photo above) on the edge of the intersecting wall and drawing a line on the top and both sides of the through wall. I mark an X with keel on the top plate to show the framers where the corners and channels will nail in place. Mark corner and channel locations precisely. A repeated small mistake can cause walls to be out of plumb once they are raised. The corner and channel marks also indicate where the double top

plates will intersect, tying together the through walls and butt walls.

Next, I mark the location of each door and window with keel. I will have made all the headers early. I lay them at their place on the plates, and then simply draw a keel line down from the header and across the two plates without using a square. Away from the header and on each plate, I mark an X at the location of the king studs that nail to the header ends. A straight line on each plate under the header indicates this space is a door or window opening (see the detail illustration on p. 199). I also mark the lengths of upper and lower cripples on the headers.

At this point, I study the plans again and detail any specials such as medicine cabinets, tub backing, openings for plumbing access, posts for beams, heating ducts, short or tall walls, or even ironing boards. Proper layout for these items before framing means you won't have to remodel a wall later.

Use a Layout Stick to Mark the Studs

The framing members in most houses are laid out on 16-in. or 24-in. centers because these measurements divide evenly into the 4×8 dimensions of standard sheathing and drywall. Maintaining a consistent layout along the length of the exterior wall is important so that panel edges always fall in the middle of a stud.

Many carpenters lay out stud locations by stretching out a tape measure, ticking the 16-in. or 24-in. increments on one plate with a pencil and returning with a square to extend the lines to both plates. This process works, but it's slow.

I use a layout stick (see the photo on the facing page) instead. This 4-ft. long aluminum bar has 1½-in. wide tabs on 16-in. and 24-in. centers and serves as a template for spacing studs. To begin a wall, I hang the first short tab on the stick ¾ in. beyond the end of the wall. This placement sets up the layout so that the centers of the studs fall on 16-in. or 24-in. centers, and sheathing edges will land centered on the studs. Then I mark both sides of all the studs for this 4-ft. section of wall, move the layout stick down, line its end up on my last mark and again mark out 4 ft. more of wall.

When I come to a door or window opening, I continue the layout, marking cripple locations on the headers. You can ease the plumber's task by laying out the studs to leave a full bay for shower and bath valves. I like to start butt-wall layout at channels, leaving a full bay's room to swing a hammer when nailing the walls together.

Because of the sheer number of marks that are required, I wander through the plated rooms and check to see whether I have missed marking a corner, a door, or even some studs here and there. An error caught at this point can save time and grief during the actual framing.

At this point, all the information necessary to frame the walls is marked on the plates. Even if I have to leave the site, a relatively inexperienced crew can have the walls standing by the end of the day without ever seeing the plans. The final step in plating comes after the studs have been nailed between the top and bottom plates. I nail on the double top plate before raising the walls. On through walls, I leave out sections of this plate at the corners and channel marks. The double top plates on the adjoining butt walls are cut one plate width longer to lap the through wall above corners and channels, tying the walls together.

Larry Haun, author of The Very Efficient Carpenter *(The Taunton Press, Inc.) and* Habitat for Humanity How to Build a House *(The Taunton Press, Inc.), has been framing houses for more than 50 years. He lives in Coos Bay, Oregon.*

Layout stick. With tabs on 16-in. and 24-in. centers, a layout stick locates studs fast. The longer teeth come into play when plates are laid side by side and are marked on their faces rather than edges.

Not-So-Rough Openings

■ BY JOHN SPIER

Although the name suggests otherwise, rough openings demand plenty of precision, especially when they are framed in load-bearing walls. Properly done, rough openings provide a place for windows and doors to fit securely, unaffected by the critical structural work being done by headers, king studs, trimmers, sills, and cripples (or sill jacks). After years of building, I've learned that getting rough openings right makes the rest of the job go smoothly.

Check the Plans First

Although the rough openings for doors and windows are specified on the plans, these dimensions are worth double-checking. It's important to note that sizes always are described width first, then height. Getting this information correct is the first step in avoiding serious frustration a few weeks down the line.

Occasionally, circumstances can require rough openings to be modified. Nonstandard floor thicknesses, specialized flashing elements, and applied sills are just a few details that can affect rough openings and should be thought through. If the building details are particularly unusual or complicated, it's smart to test the scenario as a mock-up before committing to a whole project.

Verify Rough-Opening Locations

In conventional platform framing, headers typically are sized so that the opening is at the correct height with the header tight to the top plate. Sometimes this placement needs to be modified, either by using cripples above the header or by moving the header up into the plates.

Lateral locations of rough openings usually are specified from the edges of the building to the centers of the openings, and between centers when several rough openings appear next to each other.

Before transferring layout marks to the lumber, I give the entire plan a final check. Confirm clearances, and make sure that the rough-opening layout will maintain symmetry within and between floors, if that's a priority. When there is room to move left or right, it's nice to make sure that trim details fit cleanly without ripping and squeezing.

Framing Terms

Rough opening An opening deliberately oversize by ½ in. or so for windows and doors to be shimmed plumb and level.

Header In load-bearing walls, this beam carries the load around windows and doors.

King stud Full-length studs nailed to each side of the header to support the rough-opening assembly between plates. They are the same length as the wall's common studs.

Trimmer stud (also called jack or jack stud) These studs are nailed to the king studs directly beneath the header and carry the load transferred by the header down to the bottom plate. Trimmers set the height of the rough opening.

Sill (also called bench or saddle) This 2×6 is laid flat and nailed between the trimmer studs to support the window. The sill serves as a foundation for the window's pan flashing. Adding a second sill here provides extra nailing for trim, but it is optional.

Cripple (also called sill jack) These shortened studs support the sill and act as nailing for sheathing and interior-wall finishes. They follow the same layout points as common studs for the entire wall. Nailing sill jacks to trimmers when they don't land on layout is optional.

Omit if rough opening is for a door.

Fixing Wrong Rough Openings

PROBLEM: The window opening is too short.
SOLUTION: Enlarge the opening by removing one of the two sills. If there's only one sill, cut the tops off the sill jacks to lower it.

PROBLEM: The header is too low.
SOLUTION: Substituting a smaller but stronger header works, but then the trimmers need to be replaced.

PROBLEM: The opening is too narrow.
SOLUTION: If I need to gain ¼ in. or less, I use a circular saw to shave the trimmer studs. Anything the sawblade can't get I remove with a chisel. Substituting narrower trimmers is also a good option. After checking bearing requirements, I substitute 2× trimmers with 1× or 5/4 stock. Eliminating trimmers and substituting mechanical fasteners such as Simpson HH4 header hangers (www.strongtie.com) is an excellent fix. In the case of a window opening, replace at least the top sill with a longer one.

PROBLEM: The opening is way too narrow.
SOLUTION: Leave most of the framing in place, and simply move the king stud and trimmers on one side, replacing the header and sills with longer ones.

PROBLEM: The rough opening is in the wrong location.
SOLUTION: Rather than start from scratch, I move the rough-opening assembly as a unit. I start by cutting through the nails along both top and bottom plates. If the sheathing is on already, I cut the nails with a reciprocating saw after creating some space between it and the framing with a sledgehammer. Before moving the unit, I cut the new opening in the sheathing, then install the rough-opening assembly.

Centerlines Guide the Layout Process

Most plans indicate door and window locations by noting a dimension to the centerline, so I start my layout by marking the centerlines on the wall plates. **1.** Then I measure half the width of the rough opening on both sides of the centerline. **2.** Using a Speed® Square, I finish marking the trimmer and king-stud locations.

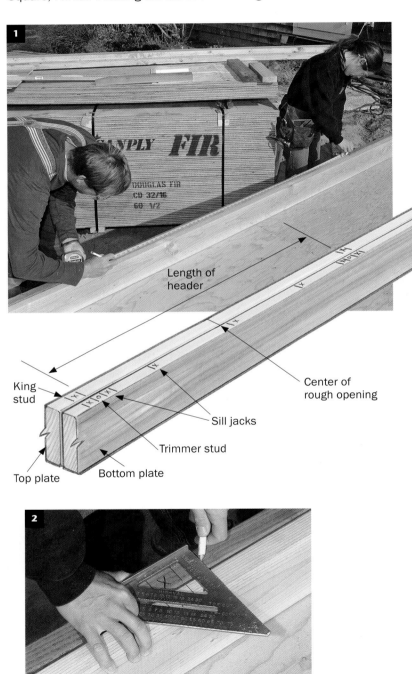

Length of header

Center of rough opening

King stud

Sill jacks

Trimmer stud

Top plate

Bottom plate

Gather All the Pieces

All the wall components that define a rough opening need to do their part in making the opening solid and square forever. The king studs on each side of the header should be straight in all directions (for example, no bow, crown, or twist). The header needs to be sized appropriately and should provide room for insulation, if possible.

Trimmers need to be continuous from the header to the bottom plate. In some areas of the country, this detail is required by code. Even if it isn't required, I still maintain that this practice is the best. Interrupting the trimmers with sills is not a good idea because the ends of the sills eventually can crush under the load. Multiple gaps, even of the slightest dimensions, can allow settling to occur. Lateral resistance of the wall is better with continuous trimmers as well.

The building code doesn't specify the numbers of trimmers, but convention (and my building inspector) requires that we have two at each end of openings over 6 ft., three over 12 ft., and stamped engineering on anything questionable.

The sills, which sometimes are called benches, need to remain flat and straight to support pan flashings under windows, and sometimes the jambs of the windows themselves. Sill jacks support the sills and maintain the stud layout of the wall for fastening sheathing and wall finishes. Some framers use a single sill and skip the sill jacks against the trimmers when they don't fall on layout; eliminating three pieces of wood here saves time and materials and still meets code. I prefer the extra nailing and solidity of double sills with end support.

> **TIP**
>
> *I like to look up rough-opening dimensions in the tables provided by window and door manufacturers. These tables are excellent, and I've learned not to second-guess them.*

Get Everything Ready

Once the location and size of rough openings are verified, I cut all the pieces before assembling the walls. This step minimizes work time and waste, and ensures a speedy and accurate process when it's time to assemble the walls.

The headers I'm using here are easy to assemble from conventional lumber. My wife, Kerri, aligns the pieces as I follow with the nail gun. We neatly stack and organize the sills, sill jacks, trimmers, and headers prior to wall framing.

Square the Wall, Then Frame the Openings

nail in the king studs and common studs first, leaving out any commons that will get in my way later when I'm nailing in the header. **1.** Then I square up the wall and tack it to the deck to keep it that way while I finish the rough openings. **2.** I install the header between the king studs next, **3.** then nail in any commons that were left out. **4.** Because doors and windows are attached to the trimmers, I nail them solidly to the king studs to minimize twisting and warping. I put two 16d nails at the top and the bottom, then one staggered every 12 in. in between. If it's a window opening, I don't nail between 10 in. and 20 in. off the floor. That way, plumbers and electricians are less likely to hit a nail when drilling. I also angle the nails slightly so that the points don't come through and damage hands and wires later.

Trimmer stud nailing pattern

Assemble the Parts in Order

While assembling the parts, it's important to keep everything flush to the inside. I keep framing tight to the deck as I nail the pieces together, and I nail carefully so that split ends don't create bumps. Maintaining a flush inside surface helps to minimize drywall cracking and also creates a smoother final wall finish.

The rough opening goes together in clearly defined steps, and following this process in order is important.Before I get into the main components of the rough opening, I nail the common and king studs through the top and the bottom plates starting from one end of the wall. For the time being, I leave out any studs or partition posts that are close enough to the king studs to prevent end-nailing the header.

Once the common and king studs are in place, I pin the bottom plate to a snapped line on the deck and square the wall. As a unit, the wall needs to be straight and square before the openings are assembled; once all the components are nailed off, any adjustment racks and bends the parts and loosens the joints.

After the wall is squared, I drop the header into the opening and nail it securely through the top plate and then through the king studs. Before I forget, I like to nail in any studs that were left out to allow for clearance of the nail gun. The trimmers are installed next, using enough nails to keep them from twisting.

If I'm framing a window opening, the next step is to nail in the sill jacks. Put the end ones in first so that they can be nailed securely to the trimmers. To get an accurate layout for the first sill, I lay it along the bottom plate, transfer the layout marks, and nail in the sill. To give drywallers and trim carpenters a little help, I keep nails in the second sill in line with the sill jacks.

Complete the Final Step

Sheathing and housewrap come next, and I usually install these materials with the framing flat on the subfloor or slab. After this, there's one final step in finishing off the rough openings in the wall.

With the wall tilted up and braced plumb, I drive a few more nails in the corners that I wasn't able to reach while the framing was flat.

John Spier is a builder on Block Island, Rhode Island. His book, Building with Engineered Lumber, *is available from The Taunton Press, Inc.*

Final detail. Once the wall is raised and braced, I finish nailing off the parts of the header that were not accessible while the wall was lying on the deck.

Cripples and Sills Complete a Window

When installing the cripples and sills for a window opening, I nail the cripples into the bottom plate first. **1.** Then I lay the sill across the cripples and transfer the layout. **2.** I nail the first sill in place, then add a second sill (entirely optional) to act as nailing for interior trim. I keep the nails in the second sill aligned with the cripples to help drywallers and trim carpenters find the framing later.

Framing Curved Walls

■ BY RYAN HAWKS

"**P**iece of cake," I replied the first time someone asked whether I could build curved walls. But as I looked over the plans, I didn't have a clue how to turn that drawing of a round room into reality. Late in the day, after the generators were silenced and the crews were on their way home or to the bar, I prowled around at a couple of jobs being built by one of the best framing contractors in San Diego.

By looking at his work, copying his notes from the floor, asking questions the next day and applying ingenuity, I built my first radius wall as if I'd done it a hundred times. That was a few years ago; I've since developed a more refined technique that I use now.

A Chalkline Represents the Base of the Radius

The job shown here is a 180-degree radius wall, or a half-circle, that extends from two parallel sidewalls (see the photo on the facing page). This radius wall is the simplest type (see the sidebar on p. 210).

To lay out the radius, you first must establish the baseline. In this case, the baseline connects the ends of the two sidewalls. I snap a chalkline on the floor to represent the baseline, then mark its center (half the length of the baseline is the outer radius).

A nail partially driven into the floor at this center mark becomes the pivot point for my tape measure, which I use as a compass to mark the inside radius of the wall plate on the floor (see the top photo on p. 210). Most tapes have a slot cut in the hook at the end for just this purpose: The edge of the nail head rides in the slot as you hold a pencil to the edge of the tape at the proper point and mark the radius to the floor (see the bottom photo on p. 210).

Complications arise when something straight, such as a door or a window, goes into a curved wall. On this job, the architect had called for the wall to be 5½ in. thick and for the door to be 3 ft. wide. If I'd built it as drawn, the center of the door-jamb head would have intruded past the wall surface. After consulting the architect, we thickened the wall to 7½ in. and downsized the door to 2 ft. 8 in. (see the illustration on p. 210)

To frame curved walls, use a tape measure as a compass and thick plywood for the wall plates.

Layout begins with a baseline and a centered nail. A chalkline between straight walls from which the curved wall will spring forms the baseline. A nail centered in this line is the centerpoint of the curve.

The slot on the hook of the tape fits around the head of the nail, allowing the tape to be used as a compass, in this case to lay out the inside radius of the wall.

Radius-Wall Variations

LARGE-RADIUS WALL PLATES

When the radius is larger than 4 ft., the centerpoint can't be on the plywood. To find the centerpoint, snap a chalkline that's at least the length of the radius on the floor. Next, bisect the plywood with a chalkline, then align the plywood's chalkline with that on the floor. Locate the centerpoint of the radius along the chalkline, and use a tape measure to swing the radius.

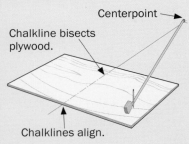

Chalkline bisects plywood.

Centerpoint

Chalklines align.

WHEN THE ARC IS LESS THAN 180 DEGREES

Jack radius walls, where the arc of the wall is less than 180 degrees, are common, but finding the radius of these walls so that you can lay them out is not so obvious. Jack radius walls are usually specified with two factors: rise and run. As an example, say the run is 8 and the rise 2. Plug these two numbers into a simple algebraic equation to get the radius (see the illustration below).

Radius = (Run² + 4Rise²) ÷ 8Rise
Radius = (8² + 4 × 2²) ÷ (8 × 2)
Radius = 5

$$\text{Radius} = (\text{Run}^2 + 4\text{Rise}^2) \div 8\text{Rise}$$
$$\text{Radius} = (8^2 + 4 \times 2^2) \div (8 \times 2)$$
$$\text{Radius} = 5$$

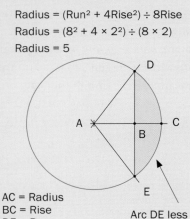

AC = Radius
BC = Rise
DE = Run

Arc DE less than 180°

Thick Plywood Makes Up the Radius-Wall Plates

I usually make radius plates from two top and two bottom layers of 1⅛-in. plywood, which adds up to the same 4½-in. thickness as three 2× plates, avoiding the need to custom-cut studs. When framing on a slab, however, as here, the bottom plate must be treated plywood, which is available in a maximum thickness of only ¾ in. Cutting six ¾-in. plates to achieve a 4½-in. plate thickness would have been a lot of work. Instead, I used two ¾-in. bottom plates and two 1⅛-in. top plates; I custom-cut the few studs that this wall required.

This particular radius was smaller than 48 in., so marking the plate radii on the plywood was straightforward. I simply set a nail in a line bisecting the sheet and used my tape as a compass to lay out the inside and outside radii on the plywood (see the top photo at right). (See the sidebar on the facing page for walls whose radius is greater than 4 ft.)

If a radius is greater than 10 ft., I can cut the plywood plates with a 7¼-in. circular saw, but I cut these tight-radius plates more conventionally using a jigsaw. After cutting the first plate, I used it as a template to mark the other plates as well as the rough sills and the nailers at the top of window and door openings.

Once the plates are cut, I place the treated bottom plate on top of the anchor bolts, lining up one end with the intersecting straight wall. I use my square to align the plate with the radius I marked on the concrete and then hammer the plate over the anchor bolts to leave indentations (see the bottom photo at right). These indentations are where I drill for the bolts.

With the bottom plate drilled, I lay it back down on the anchor bolts to check the fit and to mark the ends to be cut where it meets the next section of plate (see the left photo on p. 212). One end meets the

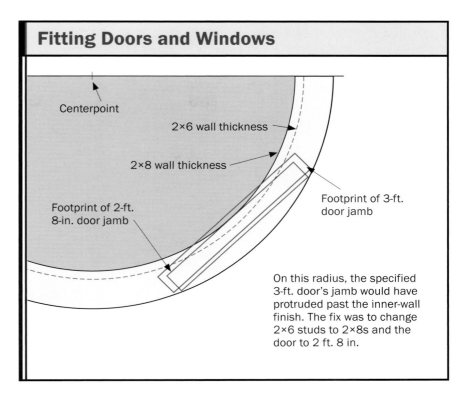

Fitting Doors and Windows

Centerpoint

2×6 wall thickness

2×8 wall thickness

Footprint of 2-ft. 8-in. door jamb

Footprint of 3-ft. door jamb

On this radius, the specified 3-ft. door's jamb would have protruded past the inner-wall finish. The fix was to change 2×6 studs to 2×8s and the door to 2 ft. 8 in.

Fit the bottom plate. The plates are marked to be cut using a tape as a compass.

Using a square to ensure that the plate aligns with the layout line, the author taps the plate over the anchor bolts to mark the hole locations.

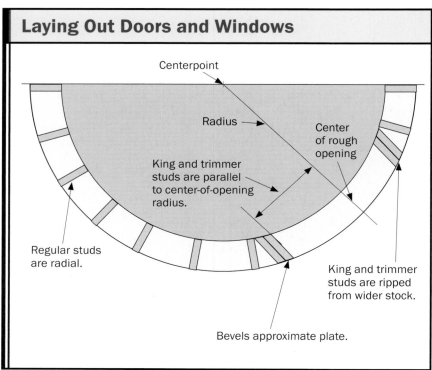

With the first section of plate drilled and placed, a square again is used to align a second section, and the author marks it for length.

Mark the rough opening. The author snaps a radius line to the center of the rough opening, then transfers that line to the bottom plate.

Laying Out Doors and Windows

Centerpoint

Radius

Center of rough opening

King and trimmer studs are parallel to center-of-opening radius.

Regular studs are radial.

King and trimmer studs are ripped from wider stock.

Bevels approximate plate.

Mark the top plate. The centerline is transferred from the bottom plate to the top (photo middle), where the locations of the kings and trimmers are marked (photo bottom).

Getting the kings and trimmers right. The simplest way to find the width and bevel of kings and trimmers is to scribe a scrap block in place (photo above) and use it to set up a tablesaw to rip the kings and trimmers from wider stock (photo left).

straight wall, and its cutline matches the baseline of the radius. If the other end is in the middle of the wall, I want this cut to fall in the center of a stud. I find this point by wrapping my tape around the outside of the plate and then marking the center of the stud that falls closest to the end of the plywood.

With the first plate section cut and in place, I repeat my actions to cut and drill remaining sections. Once the plate is in place, I detail the door and window openings on the plates (see the illustration on the facing page).

The first step in laying out the window openings is to find their centers by measuring along the outside of the plate. With the rough-opening centers marked, I snap chalklines between them and the center-point. Aligning my framing square on the chalkline, I use the square to mark the rough opening on the outside and the inside of the plate, then to connect the marks I've made (see the top right photo on the facing page). These layout lines aren't radial; rather, they're parallel to the center-of-opening radius. If they were radial, the rough opening would be narrower on the inside than on the outside.

Studs on curved walls are often spaced closer than 16 in. o.c., in this case 8 in. Laying out the outside edge of the plate ensures that the sheathing joints, though not the drywall, land on studs.

Snapping radial chalklines for each stud would ensure perfect alignment, but lining up the studs by eye works fine and is a lot faster.

The headers are standard. Only the nailers for the wall finish (extra plate stock) are curved. Assembling the headers with a square as a guide ensures their fit.

The headers are standard, straight lumber that sandwiches between two pieces of 1⅛-in. plywood, extra pieces I cut along with the plates (see the photo at left). I size this plywood by laying it on the detailed plates and marking it to fit between the king studs.

Nail the Wall Together Where There's Room to Maneuver

I like to frame these walls on a nice flat floor, but I framed this one outside because the job was a remodel with no room inside for me to frame a curved wall. Working with curved components is cumbersome, and it helps to have one person holding the wood and another one nailing.

I begin by nailing the headers to the king studs and trimmers, and then to the plates (see the middle photo on the facing page). Next come the studs, and after nailing in more than half of them, I chock the wall with a couple of scrap blocks to prevent it from rolling. The rest of the wall is filled in one stud at a time by two carpenters, one who is holding the studs and another who is doing the nailing (see the bottom left photo on the facing page).

Round walls are usually heavier than straight walls, and they like to roll. I always take extra precautions when standing a round wall; having too many carpenters raise a wall is better than not having enough. When possible, I raise radius walls from the position of having the ends of the plates, as opposed to their centers, on the floor. This approach prevents the wall from rolling as it's being raised.

Because of the existing house, that approach wasn't possible on this job. There's really no trick here: Just have more muscle than wall on hand. Because the plates differ in thickness from straight wall plates, they don't lap to tie the walls together. They're

TIP

It's possible to mark the studs exactly radial on the plates by pulling a chalkline from the centerpoint to every layout mark. I usually don't bother with this step, however, finding that I can visually check the stud alignment well enough.

Kings and Trimmers Are Custom-Cut

Because they are parallel to the centerline of their rough openings (as opposed to being radial), the king studs and trimmer studs are wider than the other studs, and their edges must be beveled to fit neatly on the plates. The bigger the rough opening, the bigger the bevel angle.

I find the rip angle and depth of the king and trimmer studs by laying a block on the plate where the king or trimmer stud sits and scribing. On this job, with its 2×8 studs, I was able to rip the kings and trimmers from 2×10 stock (see the top photos on p. 213).

After marking the rough openings, I lay out the stud locations on the outside of the plate (see the bottom photos on p. 213). Most architects note the stud spacing on the plans. Usually, the tighter the radius, the closer the stud spacing to provide backing for the wall finishes. On this job, the studs were on 8-in. centers.

simply butted, plumbed, and tied together with metal straps wrapped around both intersecting plates (see the photo at right).

Sheathing the wall is simple. I use ³⁄₈-in. CDX plywood because it's flexible, and I sheathe as I would any other wall.

Ryan Hawks is a carpenter who does side jobs in San Diego, California.

Lack of plate overlap means metal straps must tie curved wall to straight. And yes, it would be safer to perform this operation on a ladder.

Round walls are usually heavier than straight walls, and they like to roll. I always take extra precautions when standing a round wall; having too many carpenters raise a wall is better than not having enough.

Studding out. Curved walls want to roll, so filling in the studs takes two carpenters. The author starts with the kings and trimmers (photo above left). Next come the regular studs (photo left) and the midwall blocking, which he scribes from the laid-out plates (photo above).

Framing Big Gable Walls

■ BY LYNN HAYWARD

Framing quickly and efficiently is all about doing things for a reason rather than doing things just to see wood go up fast. Many builders frame the roof first, then frame the gable walls under it. Although they seem to be getting a lot done fast (because the roof is up), it's actually very dangerous to dangle off a ladder while maneuvering full sheets of plywood and a nail gun, trying to sheathe the gable wall that just went up so quickly.

In the long run, I find that building the whole end of the house in place on the floor is much safer and more efficient. If the gable end has any kind of unusual trim detailing, as was the case on the house featured here, building flat on the deck makes even more sense.

Don't Snap Lines for Every Stud

Even builders who frame gable walls on the deck often expend more time and effort than they need to. A common practice is to snap the entire gable-wall layout on the deck: individual studs, top and bottom plates, and rafter layout.

You don't need to work this way, however. The only chalkline I snap on the deck is the line representing the inside edge of the bottom plate. Because I use the rafters to define the top of the wall (see the top photo on p. 218), any additional lines are superfluous.

My single chalkline is snapped 5½ in. in from the outside of the framing (for

Build gable walls on the floor, lift them with specialized jacks, and eliminate extra footsteps wherever possible.

Build the Outside First

To sheathe a wall before standing it, the wall should be straight and square. My process begins at the bottom plate and ends at the rafters, which define the top of the wall. Because the wall in the photo at right gets a large French door, I omitted the central section of the bottom plate rather than cutting it out later. When doing this, it's important to toenail the plate in its exact location (see the photo below) to keep the rough opening plumb and square.

1. TOENAIL THE BOTTOM PLATE TO THE SUBFLOOR

Snap a chalkline 5½ in. from the floor framing. To locate the line, use a framing square to plane up from the rim joist instead of hooking the subfloor, which may not be cut straight. Many framers toenail the bottom plate from the inside so that the toenail is easier to remove after the wall is raised. The reason why it's easier to remove the nail is exactly the reason why I do it differently: The nail pulls out as the wall is stood. If you toenail from the outside, the nail stays in the deck, which helps to prevent the wall from slipping out as it's raised. I go a step farther and nail metal straps to the wall and floor before standing the wall. After the wall is raised, it's fairly easy to pull the nails and remove the exposed portion of the straps.

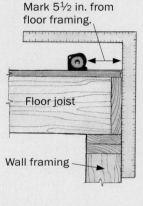

Mark 5½ in. from floor framing.

Floor joist

Wall framing

16d toenail

Metal straps

Toenail from the outside.

Toenail end of wall at rough-opening location.

2. SET THE END POSTS OF THE WALL

If the gable intersects a kneewall, as pictured, the post height equals the kneewall height minus the bottom plate.

Continuous bottom plate if no rough opening

Kneewall posts

4-in. blocks keep rafters flush with wall framing.

3. SET THE RAFTERS IN PLACE

Use 4-in. spacer blocks to hold the rafters off the floor and flush to the face of the kneewall posts. Between the rafters' plumb cuts, set a 2×6 where the ridge board will sit.

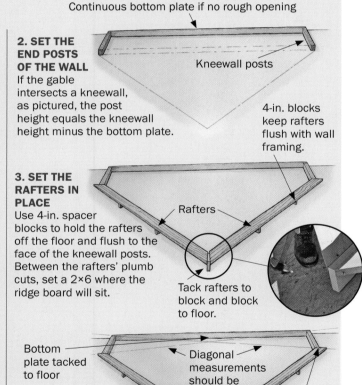

Rafters

Tack rafters to block and block to floor.

Bottom plate tacked to floor

Diagonal measurements should be equal.

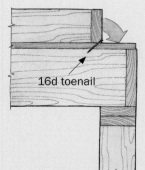

4. SQUARE THE WALL

To get accurate stud lengths, the outside of the wall has to be square. With the bottom and top tacked to the floor, the kneewall posts can be adjusted.

Top of wall tacked to floor

Adjust diagonal measurements by moving the top of the kneewall posts.

2×6 walls). Rather than hooking the subfloor and then measuring in, I use a straightedge, such as a framing square, to plane up from the framing. Following this procedure keeps the outside walls consistently straight from one floor to the next.

I cut the bottom plate to length and toe-nail it to the deck through the outer face of the plate. The nail acts as a hinge when my crew and I are lifting the wall. If you toenail through the inside face, the nail will pull out of the deck and let the wall slide bottom first as you raise it. I also use short lengths of metal strapping nailed to the floor and to the bottom plate to prevent the bottom of the wall from kicking out during lifting. After the wall is in place, we pull these hinge nails and the nails in the straps with a cat's paw, then cut off the metal straps.

Rafters Replace the Top Plates

We eliminate the top plates in gable walls in favor of notching the studs to accept the rafters. This method makes stronger and lighter walls, and it uses less wood. We nail the kneewall posts at a right angle to the ends of the bottom plate and tack them to the floor to hold them in place while we build the gable wall. Atop the kneewall posts, we place the straightest two rafters in the pile and nail them down. At the peak, we tack a 2×6 block between the plumb cuts in the rafters with its bottom placed exactly where the bottom of the ridge board will be. A stud will be cut to fit below this block, and the block will be removed before roof framing. This puts the ridge at exactly the right height during roof framing.

A Stronger Design and a Faster Layout

Some framers build gable walls the same way they frame typical sidewalls: with a bottom plate, a double top plate, and studs in between. I don't do it this way. Instead, I skip the top plates and notch the studs to accept a rafter. This approach is stronger because it eliminates a hinge point by nailing through the face of the rafter into the stud. This method also uses less wood and is faster to assemble.

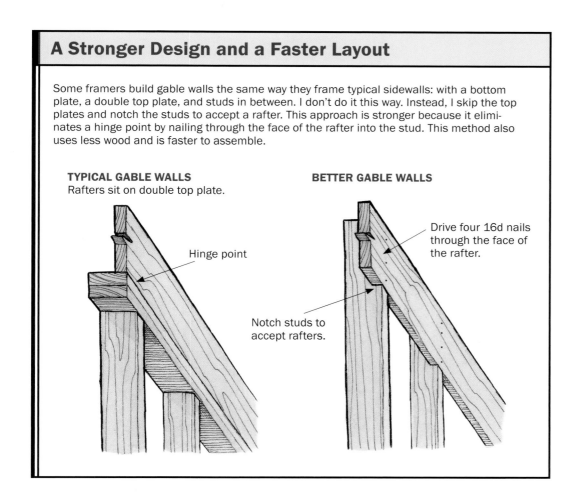

TYPICAL GABLE WALLS
Rafters sit on double top plate.

Hinge point

BETTER GABLE WALLS

Drive four 16d nails through the face of the rafter.

Notch studs to accept rafters.

Lay Out Studs with a Framing Square

With stair buttons attached to the tongue and blade of a framing square, you can mark the gable wall's stud layout right on the face of each rafter. Button position depends on roof pitch. The 1½-in.-wide tongue of the square should remain perpendicular to the wall's bottom plate as you move up the rafter, stepping off a 16-in. run. Tracing against the edges of the tongue gives you the placement of the stud.

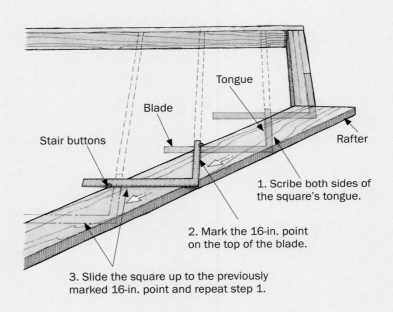

Tongue

Blade

Stair buttons

Rafter

1. Scribe both sides of the square's tongue.

2. Mark the 16-in. point on the top of the blade.

3. Slide the square up to the previously marked 16-in. point and repeat step 1.

With the outline of the wall nailed together, and the bottom plate and the ridge block tacked to the subfloor, we square the wall using the kneewall portion as our rectangle for equalizing the diagonals. When the wall is perfectly square, we tack the end posts in place to the floor and turn to framing the inside of the wall.

Mark Stud Layout on the Rafters with a Framing Square

After laying out the wall on the bottom plate, we transfer the wall layout to the face of each rafter using a framing square and a pair of screw-on stair buttons (see the sidebar above). In stair-layout language, you have to position the buttons on the square so that the broader blade steps off a

Once the stud layout is marked on the rafters (see the sidebar on the facing page), the lengths of all gable studs can be measured and studs can be cut all at once to save time. Measure studs from the wall's bottom plate to the long point of the notch at the bottom edge of the rafter. Once the first notch is cut, use the offcut to mark the notch angle on all remaining studs.

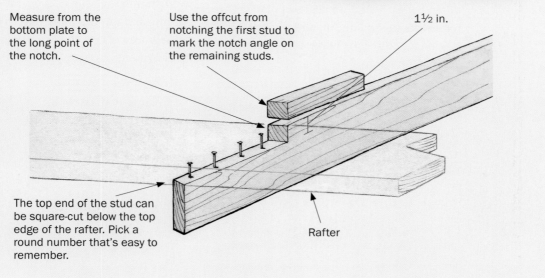

Measure from the bottom plate to the long point of the notch.

Use the offcut from notching the first stud to mark the notch angle on the remaining studs.

1½ in.

The top end of the stud can be square-cut below the top edge of the rafter. Pick a round number that's easy to remember.

Rafter

16-in. run, while the square's tongue remains plumb—or perpendicular to the bottom plate, because the wall is framed on the deck.

Before measuring the studs, we stretch a gauge string along the top of the rafter to see if there's a bow. If we find a bow, we cut the middle stud short by a little more than what the string shows. We then nail the stud on layout to the bottom plate, tack it to the floor with a toenail, and use a block (which is nailed to the subfloor) and a short length of 2×4 to pry the rafter straight.

We face-nail the rafter to the stud and then release the lever. This step usually takes care of any discrepancies.

Cut All the Studs at Once, Not One Stud at a Time

To get the stud length, I hook my tape on a scrap of 2×6 tight against the bottom plate and measure to the long point of the beveled notch. It's better to measure to the long point because after cutting the beveled notch, you may need to hook your tape on the long point and pull the overall length.

As I measure the studs, I mark their lengths on the face of the rafter, then on a block of scrap wood. This way, I can cut all the studs necessary, or at least six or eight at a time. Because the notch in the stud is beveled, I need to mark the slope of the rafter on each stud, but I do this only once (with a Speed Square). For each subsequent cut, I use the offcut from the first stud as a scribing block.

You should be able to cut two gable-wall studs from each 16-ft. 2×6: a long and a short piece, or two medium pieces. The top of the stud is cut square below the top of the rafter. For each stud, I make two marks: One represents the square cut, and the other is the long point of the bevel. I trace the

1½-in. width of the notch with the tongue of the framing square, scribe, then cut the notch. As I cut the studs, I cross the lengths off the list.

Use a Router to Cut Out Windows, Doors, and Sheathing Edges

The framing is straightforward. Check the rafter against the gauge string and correct if necessary. Frame window and door openings in the center of the wall first, then work your way out. Finally, face-nail the rafter to the studs.

I begin sheathing by snapping a chalkline about 3 ft. up from the bottom plate. This overhanging sheathing covers the floor framing and ties the gable wall to the rest of the house. To get the exact overhang measurement, measure down from the

Smooth framing strategy. Before you start filling in the wall with studs, run a string along the top of the rafter to check for straightness. If there's a substantial bow, bend the rafter back straight with a lever and a block nailed to the subfloor. Then secure it with a stud or two.

Install a dummy ridge board. A short 2×6 provides the proper spacing for a ridge board to fit between rafters after the wall is raised.

Frame from the inside out. If you frame the window openings first, you'll be able to nail sideways into the header with a nail gun for a strong connection. Cripple studs above the header are placed last because they need to be measured after the window opening is framed.

subfloor to the top of the first floor's wall sheathing. Subtract ⅛ in. from this number. It's important that the sheathing not hang down too far, or it will prevent the gable wall from resting on the subfloor, making it unstable.

As the framers sheathe the wall, I break out my 3-hp router with a panel-cutting pilot bit, and I cut out the window and door openings as well as the overhang along the top of the rafters. This router technique is much faster than marking, measuring, and snapping a bunch of lines to represent these cutouts, and then using a circular saw to make the cuts. This technique is also a lot cleaner, leaving no sheathing bumps to hinder the window and door installation.

When installing the housewrap, we make sure to snap a line, similar to the one for sheathing, to allow overlap on the first floor. We also allow for about a foot of overlap at

Wall sheathing spans the rim joist. Hanging the sheathing below the bottom of the wall ties the two stories together structurally. Don't hang it down too far, though, or it will prevent the wall from sitting on the floor fully.

Cut openings quickly with a panel-cutting bit. I look for a plunge-cutting panel bit with a cutting length of 1⅛ in. or 1¼ in. This allows me to get twice the life out of the bit: After I dull the top half, I reset the bit lower for a fresh cutting surface. Porter-Cable® makes one for around $16.

the corners, stapling along the corner and folding back the extra. As we progress along the wall, we pull the housewrap tight. It's important to keep the housewrap tight and to avoid big bubbles that can cause a problem when snapping lines for the clapboards or the siding.

I don't like to install windows or apply siding to a wall before standing it up because the racking that takes place during the lifting process can loosen the nails. I do, however, apply exterior trim to the gable. I make sure the trim has been painted with at least two coats of exterior paint. Here's a tip about the gable fascia: As wood dries, it shrinks. Because the board shrinks equally from side to side, it shrinks the most along the short point of the miter, the widest part of the board. To minimize this problem, we cut the fascia so that it has a small gap at the long point. A couple of weeks later, after the short points have shrunk, the gap is

uniform. This usually coincides with the end of the roof dry-in, and as long as we're up there, we toenail the gap together with a stainless-steel siding nail for a tight, long-lasting joint.

Wall Jacks Make Easy Work of Lifting Heavy Walls

After the wall is framed and sheathed with housewrap and the fascia is applied, it weighs a lot. I use wall lifts made by Proctor® (www.proctorp.com; about $875* for 16-ft. jacks). They're basically steel pipes with a come-along mounted to them (they telescope for easy storage). At the bottom of the rod is a hinged shoe; one leaf of the hinge is nailed to the subfloor and the other leaf is welded to the pole.

The pole can move with the wall as the wall goes up. The pole starts out standing basically plumb, and as the wall is jacked up, the top moves farther and farther away from the shoe, hence the hinge. At the top of the pole is a pulley through which the cable from the come-along runs. At the end of the cable is a bracket nailed into the top of the wall.

The wall jacks also have a small bracket at the top of the pole to stop the wall from falling past the 90-degree plumb position and falling off the house. After the wall is standing plumb, we attach braces to hold it that way, and we remove the wall jacks.

With the wall stood and braced, we pull the toenails and then cut the straps; then we nail the wall to the line. This sometimes takes a little persuasion.

Prices are from 2006.

Lynn Hayward *owns and operates Hayward-Robertson Builders in Northport, Maine.*

A little persuasion gets the wall back on the line. With the wall stood and braced, we nail down the bottom plate, adjusting it to the line as we go. We also drive a few 8d nails through the sheathing into the rim joist to squeeze it tight before nailing off the sheathing with a nail gun.

Lift with Your Head, Not with Your Back

Lifting exterior walls with wall jacks is safer in the short and long run: The wall doesn't fall on anyone during lifting, and my back lasts longer. The 16-ft. jacks (about $875) consist of two telescoping steel poles, a come-along, and a hinged shoe. A bracket at the end of the come-along cable is nailed into the top of the wall, the hinged shoe is nailed into the subfloor, and the wall is raised by cranking the come-along. A hook on top of the poles keeps the wall from falling off the house after it has been stood.

Raising a Gable Wall

■ BY JOHN SPIER

There is no easy way to set up the staging for building the gable wall of a house. Most approaches involve pump jacks, wall brackets, an assortment of ladders, and too much time. Even a simple gable can be tough to stage, and I've seen gables with their peaks 40 ft. off the ground and others that had to be built above another roof that were nearly impossible. One of the scariest staging jobs that I ever saw was for a gable that was built over a large greenhouse.

Fortunately, there is a way to avoid staging altogether: If you build the gable walls on the deck, you can raise them into place (see the photo on the facing page). Often, when I raise a gable wall, it has been framed, sheathed, sided, trimmed, and painted. The closest anyone has to get to that gable again is when the roofer looks over the edge.

The Gable Is Laid Out on the Deck

In some cases the gable walls that I build and raise include a full-height wall or a kneewall below the triangular portion of the gable. For the Cape-style home shown here, the top floor was to be under a simple gable roof with dormers on both sides, so we built

and raised only the triangular section of the gable wall.

After I've built the uppermost floor of the house, I snap chalklines for the perimeter walls. If the floor isn't perfectly square, I make a few minor adjustments to get the sides parallel and the ends square, a procedure that will simplify the roof framing immeasurably. The next step is to snap lines representing the peaks of both gable-end walls.

I start by snapping the centerline, or ridgeline of the house, and the tops of the sidewalls, if any. Then, using the distance from the centerline to the sides, along with the roof pitch, I calculate the height of the peak and snap lines for the top of the gable wall (see the top photo on the facing page). These lines represent the bottoms of the rafters, which are also the top-plate lines. Even if the two gable-end layouts overlap in the middle of the deck, I snap them. As a final check, I measure the four top-plate lines. If they're not all the same, I figure out why and fix the problem.

At this point, I use the length of the top plate to lay out and cut a pattern rafter. I check it for fit by laying it down in place on the gable-end chalklines. I then make four rafters for the gable walls, label the pattern clearly, and set it aside for framing the rest of the roof.

Finding the peak. The first step is laying out the upper plates and snapping chalklines for other framing details.

Framing the entire gable while it's flat on the deck is fast and easy. Raising it is a challenge and a danger.

Blocks keep the plates straight. Top plates are tacked to 2× blocks that have been ripped to 2 in. The blocks will also support the gable rafters.

Steel Bands Act as Hinges

Before I start assembling the wall, I lay out a few more things on the floor. I locate any interior walls that intersect with the gable so that I can include partition posts. Any other items that require a nailer, such as a tub or shower or closet shelves, are also located and marked. Finally, I locate the windows and snap lines on the floor at the insides of the king studs. Now I'm ready to start framing.

First, I cut and assemble the plates. To hold the plates straight, I use short 2× blocks ripped to 2 in. for 2×4 plates and 4 in. for 2×6 plates and nail them to the deck every 6 ft. or so to hold the plates straight. The blocks are nailed to the floor on edge above the plates, where they serve double duty by

Metal straps act as wall hinges. Nailed to the floor framing and then to the bottom plates, straps keep the wall from slipping off the deck's edge.

temporarily supporting the gable rafters. The plates are then tacked to the blocks (see the top photo on the facing page).

For the bottom plates, I use 12-in. ridge ties or short lengths of metal banding that act as hinges when the wall is lifted (see the bottom photo on the facing page). The bands are nailed through the floor sheathing and into a joist, block, or other framing below. The plates are set on top of the steel band, and the band is bent up and nailed to what will be the underside of the plate. These "hinges" are essential to stop the bottom of the wall from kicking out during the raising. After the wall is in position, I cut off the inner part of the band with a reciprocating saw.

Next, I install the king studs for the windows and add the trimmers, headers, sills, sill jacks, and cripples. If there is no window in the center, I install a full-length center stud or post to carry the ridge. If there is a center window, I put in the ridge post above the header. The studs are then laid out and installed for the gable wall using the same spacing that I established on the first floor, which keeps the framing neatly in line from foundation to peak (see the photo below). After the studs are cut and nailed in, I add blocking, fire-stops, nailers, and any other framing that might need to go in the wall. If there are gable louvers or other vents, they are framed in, too.

The next step is adding the rafters that sit on the gable-end plates. I lay the rafters on top of the blocks holding the plates straight and nail the rafters to the plates. The top ends of the rafters get shaved slightly so that

Bring on the sheathing. The completed gable frame— including rafters, studs, and blocking—lies on the deck, ready for sheathing. Horizontal edge-blocking for the sheathing is required for this high-wind area.

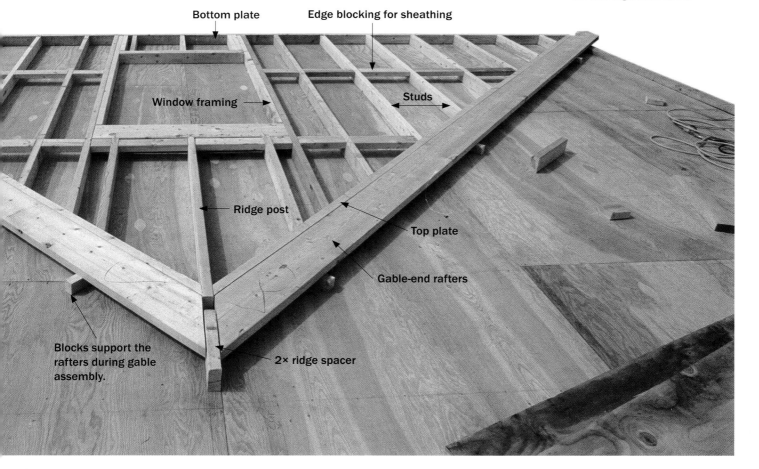

Bottom plate

Edge blocking for sheathing

Window framing

Studs

Ridge post

Top plate

Gable-end rafters

Blocks support the rafters during gable assembly.

2× ridge spacer

the ridge will slide in easily after the walls are up. I also make sure the bird's mouths will fit the sidewalls after the gables are raised, and I pull the nails holding the plates to the blocks.

Add the Sheathing, Housewrap, Trim, and Shingles

With the framing complete, I turn my attention to sheathing the wall (see the top right photo below). I leave the appropriate amount of sheathing extending beyond the bottom plate to meet the sheathing of the wall below, subtracting ½ in. from my exact measurement to allow for shrinkage, settling, compression of the floor, and carpenter error. If there is a lot of loose ply-

wood hanging out beyond the edge of the house, I mark it so that no one walks out and falls through.

I finish sheathing the entire gable and cut out window openings. As you might expect, nailing off the sheathing while the wall is lying flat gives us a big advantage in terms of time, effort, and accuracy. The housewrap also goes on much quicker and easier while the wall is horizontal (see the bottom right photo below).

After the housewrap, I nail on the felt-paper splines for the windows. Another strip of felt paper goes along the top of the wall, and sometimes a piece is needed for a box return or other trim detail.

Now I'm ready to tackle the rake detail. If there is an overhanging rake, I use the pattern rafter to cut the fly rafters, adding half the ridge thickness to each so that they meet

Fitting gable trim in comfort. Instead of working off tall ladders or awkward staging to get all the trim cuts right, the author fits all the trim perfectly with the wall lying flat.

It sure beats carrying plywood up an extension ladder. Sheathing the gable wall while it's flat means that sheathing doesn't have to be held in place while it's being nailed. Cutting and nailing also go much quicker horizontally.

Gift-wrapped gable. With the wall lying flat, the housewrap can be stretched out and stapled without wrinkles.

Gable Vents for High-Wind Areas

The storms that buffet Block Island are typically accompanied by gale-force winds blowing in off the water. Precipitation in these storms travels horizontally, forcing its way into even the smallest openings. Traditional louvered gable vents inevitably leak.

The bow vent (see the photo on p. 233) is a decorative but functional solution. This vertical eyebrow arrangement admits air for ventilation but shields the opening from rain or snow. Local codes may vary in determining vent size, but for this job, we cut 2-in. by 24-in. openings on both gables (see the top photo at right). We then staple copper or bronze screen over the opening. Two bow-shaped nailers are then cut out of pressure-treated or cedar 2× stock—one large bow about 2 ft. wider than the opening and another proportionately smaller bow, both with slight angles on the tops for the angle of the shingles. The center of the larger nailer is cut out, leaving a hole the same size as the vent opening. The solid upper nailer stiffens the shingles and limits the space in the vent to discourage birds and insects. The nailers are then attached to the gable wall above and below the opening (see the middle photo at right). After shingling up to the lower 2× nailer, we shingle over both nailers starting in the middle and working to the sides (see the bottom photo at right). A starter course under the bottom course ensures a weathertight vent. We continue shingling to the peak, blending bow shingles with wall shingles.

These vents wc k best in conjunction with a small attic space, soffit vents, and vent baffles. When correctly sized and built, bow vents provide adequate ventilation even on fair-weather days.

A hole for air but not water. A bow vent begins with a hole cut in the wall properly sized for the area to be vented.

The bow is framed with 2×s. Beveled nailers are attached above and below the vent opening.

Shingles follow the line of the bow. Shingles are nailed to the nailers using a stretched string as a guide.

Usually about six or eight people are required to raise heavy gable walls. Of course, a crane or commercially available wall jacks could make the job easier, but I don't happen to have either. The trick is not to raise a really big or heavy wall all in one motion.

in the middle. The fly rafters are installed on lookout blocks, usually at 24 in. o.c. A framing square on the floor helps us to keep the overhang in the proper roof plane during assembly. There was no gable overhang with this project, so we simply nailed a subrake on top of the felt-paper strips.

At this point, we can install most of the trim. If there are cornice returns, we build them and start our trim there. If not, we just leave the rake boards long to be cut and fitted when the rest of the roof is framed and trimmed (see the left photo on p. 230).

To save weight for the lift, we usually leave out the windows. A course line for the shingles is established above the finished window height, and we shingle up from there. (For smaller gables, we install the windows and shingle the entire wall.) We nail the lowest course near the tops of the shingles so that the courses below can be slipped in underneath. (An inconspicuous row of stainless-steel nails will secure them later.) We also build our bow vents at this time (see the sidebar on p. 231). Finally, if time and weather permit, we fill nail holes, caulk, and put three coats of paint on all the trim.

Braces help with the lift. With the wall held in place at about a 45-degree angle, 2× braces are nailed high up on the studs (photo above). The braces then help to control the final push (photo at right) and finally are nailed to blocks attached to the floor framing.

The Gables Are Raised in Stages

Some of the heaviest gables we've ever stood up were 30 ft. long and 16 ft. high, and framed with 2×6s. Usually about six or eight people are required to raise walls such as these. Of course, a crane or commercially available wall jacks could make the job easier, but I don't happen to have either. The trick is not to raise a really big or heavy wall all in one motion.

We start with a couple of strong sawhorses standing by and lift the peak enough to kick the horses underneath. For this stage, everyone is lifting close to the peak for

maximum leverage. With the wall resting safely on the horses, we next get a couple of 2×6 studs ready for props and lift the wall enough to prop it at the window headers, usually about 45 degrees. The gables shown here were small enough to be lifted directly to the 45-degree position.

At this point, the heaviest lifting is done, and control becomes our biggest concern. We nail in two or three braces near the top of the wall (see the photo on the facing page). The braces have to be long enough so that they will be at a 45-degree angle with the wall standing plumb. By angling several spikes through the brace and into a stud, a strong attachment is made that still allows the brace to pivot as the wall is raised.

The final lift is made using the braces to help push and control the top of the wall (see the photo below). Once it is upright and has been plumbed, the braces are nailed off to blocks nailed through the floor sheathing and into the joists. Sometimes we add a few additional braces to keep the rakes straight

Tying the gable to the house. Once the gable is braced in position, the bottom plate is nailed off (photo right) and the sheathing that was left overhanging is nailed to the house framing (photo below).

or for peace of mind overnight. When the wall is up and braced, the bottom plate and the overlapping part of the sheathing can be nailed off (see the photos above and left), tying the gable to the rest of the house.

A Couple of Additional Tips

If the peaks of the gable walls overlap by a little in the middle of the floor, we often build the first gable and lift it onto the sawhorses so that it is out of the way for building the second. This method allows both gables to be painted at once and means assembling extra hands for just one big lifting party.

Keep an eye on the wind while you're lifting the gables. If you're lifting into the wind, it's no problem. Just don't get squashed if you lose control of the wall. If the wind is behind you pushing the wall up and out, secure a safety rope to stop the wall before it goes beyond plumb.

Last, when the siding and painting subs show up, drive a hard bargain. The toughest part of their jobs has already been done.

John Spier is a builder on Block Island, Rhode Island. His book, Building with Engineered Lumber, *is available from The Taunton Press.*

Better Framing with Factory- Built Walls

■ BY FERNANDO PAGÉS RUIZ

Before I started building houses with factory-made walls, my five-man crew needed two weeks to frame a house. Today, the job gets done in just five days, with a crew of three.

Building with factory-framed walls demands good planning and an organized approach to the work that is done on site, which I'll discuss ahead. Once you make these adjustments, you'll be surprised by the benefits you discover (see the sidebar on p. 236). It wasn't easy persuading my crew to switch to factory-built walls, but now they wouldn't think of building any other way.

Factory-built wall panels can be ordered in different ways. You can get them framed with or without sheathing; you can have the siding installed; you can even have wall

Delivered to your site, prefabricated panels can save you time and improve quality on production houses or a custom house.

Seven Reasons to Let Someone Else Frame the Walls

1. A smaller crew

One skilled framer and two helpers can erect factory-built walls faster than a five-member crew could build them.

2. A faster job

An average house constructed with factory-built walls can be fully framed in five or fewer days.

3. The quality is unbeatable

Every stud is cut with radial-arm-saw accuracy. In the factory, it's easy for workers to frame rough openings precisely and to nail every square inch of sheathing without missing a stud.

4. Weather is not a concern

Rainy days don't slow production, and the controlled environment of the factory means framing lumber and sheathing stay clean and dry throughout the process.

5. Prices are stable

Because the factory builds so many houses, the manufacturer can order lumber and pass along some of the savings to builders.

6. Minimal waste

Offcuts are minimal; so is the amount of unusable lumber. As a result, the site stays clean, and your waste-disposal expenses are trimmed significantly.

7. Special details are no problem

Before the factory builds the walls, you can edit the computer-generated plans. It's easy to eliminate unnecessary studs, add blocking for towel racks or cabinets, or make other framing alterations.

panels delivered with the windows installed. But windows break, and finished surfaces like siding can be damaged easily. So I build all my houses using factory-framed and sheathed panels for exterior walls, and factory-framed (no sheathing) panels for interior walls. Wall panels can be fabricated to any length, but they typically come in 12-ft. to 16-ft. lengths that a small crew can handle easily.

Computers Engineer the Walls

Building with factory-made walls is a lot like building with roof trusses. Engineers at the plant plot your plans into a computer, and unless you specify otherwise, their software automatically calculates the engineering values for headers, posts, and shear walls. The resulting wall-by-wall drawings provide a distinct advantage over conventional framing plans because you can use them to edit the framing with unprecedented detail.

I build the same house plans over and over again. For me, the advantage of this system is that I can incorporate what I learn during the construction of one house into the plans for the next house. For example, if the plumber says the framing over a sink is blocking a vent, I can make a note and forward the change to the factory. But even on a custom home, you can look at the factory's framing plans before they build walls and specify detailed information like hold-down locations, complicated shear-wall nailing, or blocking for kitchen cabinets. The plant's computer then makes sure all these details are framed in the right places.

Plan Ahead to Get Walls on Time

The only drawback and the greatest learning curve with factory-framed walls is ordering the walls and timing their delivery. You can't just call the plant and get walls at a moment's notice. You have to requisition walls in advance to get them delivered on time.

Some building contractors wait until the foundation has been poured before ordering walls. The factory then can send a representative to measure the foundation and adjust the plan dimensions accordingly. If you can afford the downtime, this is the safest way to do the ordering.

I don't like to add idle time to my construction schedule, so I order walls before breaking ground. The factory builds according to my house plans without verifying any of the job-site measurements. This approach means I have to make sure the foundation is perfect. But the extra effort is worthwhile because the walls arrive on the job the day after the slab goes in. Of course, if the house has a basement or crawlspace foundation, I don't want the walls to arrive until the first-floor deck has been framed and sheathed.

If you give two weeks' notice, you should have walls on schedule. But keep in mind that your framer must be just as punctual. When the walls arrive, someone must be there to unload and organize them. It's important to make sure each wall section is placed in a convenient spot.

It's Up to the Framers to Set the Panels Square, Plumb, and Level

When the delivery truck arrives, framers have to unload the walls in sequence and stack them in order of use: interior walls in the center, exterior walls at the perimeter, and second-story walls on blocks out of the way. The placement of second-story walls is critical because there's little room to move

> *The only drawback and the greatest learning curve with factory-framed walls is ordering the walls and timing their delivery.*

around large components. Factory components come labeled, so all you have to do is refer to the factory's plans to find out which walls go where.

Just as in conventional framing, you start by snapping lines on the foundation or the floor deck. You then stand up perimeter walls, nail off double plates, and install temporary braces. Next, you erect the interior walls, taking care not to box a panel section into a room, kind of like not painting yourself into a corner.

The most critical step is lining up the top plates, then squaring and straightening the walls (see the photos at right). Because the panels typically come in 12-ft. or 16-ft. sections, you have to mate several to create a long wall. To keep the panels level, the top plates must line up flush before you nail two walls together. Sometimes you need to shim the bottom plate of one wall to level it with the adjacent panel.

One of the nice things about factory walls is that the top plates come marked at every intersection. This makes light work of measuring and fitting the double plate while leaving cutouts for intersecting walls. Most of the double plates come preassembled, but you have to cut and fit plates over wall joints.

On exterior panel joints and corners, the sheathing overlays the adjacent panel to create a structural splice. It's important to remember to nail the sheathing on these corners and laps. My framers nail off every panel before moving to the next wall section.

Once all the perimeter walls are up, it's important to square the house and align the plates with a string. After the walls are plumb and true, we brace them with 2×4s and continue with the second floor.

Setting Wall Panels

First, get adjoining panels level, flush, and plumb. Then brace each section securely.

1

Shim the bottom plate. To correct imperfections in the floor framing, drive a wedge under the bottom plate. When the top plates are level, nail the wall sections together and remove the shim.

2

Nail the end studs together to join wall sections. When the top plates in adjoining panels are level and flush, nail the end studs together with 16d nails.

3

Add the double plate over wall intersections. Make sure to bridge the intersection of two wall panels with a double top plate. The top plates are marked for intersecting walls (inset photo).

4

Nail off the intersections from outside. The wall sheathing extends over intersections and corners to create a structural splice. Nail the sections together from the outside.

Plumb and brace each wall section. As you set the wall sections, check them for plumb and temporarily brace them with a 2×4 every 8 ft. to 12 ft.

5

Three Common Mistakes and How to Avoid Them

Reverse plan orientation

If you forget to tell the panel factory that you have decided to reverse the floor plan (garage on the left instead of on the right), you have just committed a common error that can't be fixed inexpensively. The walls will be framed backward and inside out, with the sheathing on the interior face of the exterior walls. Once you finish fixing this "little" oversight by flipping the walls, reframing the openings, and stripping, then reinstalling the exterior sheathing, you could have framed the house twice.

Changes and options

If you send the factory your plans and then negotiate changes with your customer, don't be surprised to discover that the factory already has built your walls by the time you submit the modifications. If the changes are minor, it's not difficult to remodel walls on site, although you defeat the advantages of using factory-built walls.

Foundation fudges

Before the factory-built walls are delivered, take the time to check the foundation for square and level. It's easier to deal with dimension and squaring mistakes than the ills of poor leveling. You can fix slightly out-of-square foundations by building the floor deck slightly oversize in both directions. This allows you to square off the deck and lose the foundation error. And you can add plywood strips as spacers between wall sections to gain an inch or two.

A foundation that's not level presents a bigger problem because the walls come off the assembly line square and true. Exterior walls won't conform to dips in the slab or stemwalls. You'll have to fix these problems by shimming the mudsill.

Are Factory-Built Walls the Future of Framing?

Just as roof-truss manufacturers have established themselves by providing one-stop engineering and production services, wall manufacturers can deliver an integrated structural package. This proves most beneficial to builders in earthquake and hurricane regions, where pre-engineered shear-wall systems can speed up job-site construction and reduce the need for elaborate hold-down systems.

Hardware companies, wall manufacturers, and truss plants have teamed up to develop a whole-house systems approach to framing, which promises to coordinate the delivery of a pre-engineered component system suitable to any hazard area from Northern California to Southern Florida.

I advocate using advanced framing techniques that reduce lumber consumption. One obstacle to implementing such techniques is training framers on the subtleties of optimized structural framing and energy-efficient detailing. Because the panel factory provides custom-engineering services, I can design walls using these advanced techniques.

One day, factory-built walls may become as conventional as roof trusses. Until then, I thrive on the competitive advantage that wall panels provide me.

Fernando Pagés Ruiz, a frequent contributor to Fine Homebuilding, *operates Brighton Construction Company in Lincoln, Nebraska.*

Curved Ceiling? No Problem

■ BY MICHAEL CHANDLER

A cathedral ceiling can open up a room dramatically, but if the ceiling is framed with a massive structural ridge beam, the beam will be a challenge to hide. An obvious solution is either to use bigger rafters or to fur down the ceiling to hide the ridge. Oversize rafters are a waste of wood, so my three-person crew opts for furring. As long as we're installing furring, why not have fun and curve the ceiling? The furring turns a chore into a delightful, economical upgrade. Adding 1×4 furring to the framing package costs less than increasing the rafters to 2×12s or I-joists. We can install the backing and the curved furring on a 27-ft. by 27-ft. ceiling in about three hours.

Strike a Curve, and Locate Backing

We use 1×4 #2 spruce furring. It's available in lengths up to only 16 ft., so most jobs require more than one piece to span the ceiling. To control the curve of the 1×4s and to support the ceiling, we fasten backing boards across the rafters, spacing them 3 ft. to 4 ft. apart. Just as when installing strongbacks, different combinations of dimensional lumber can be used for backing, depending on the offset required from the bottom edge of the rafter to back up the 1×4 curve.

On the project shown here, we bent a 16-ft. 1×4 between the end wall and the ridge beam, and simply traced the resulting curve on the gable-end drywall. Measuring down from the rafter to the curve gave us the distance that the backing boards would need to span.

As shown on pp. 242–243, the resulting curve isn't a true arc, but a curve with flattened ends. We add fire-blocking in the walls where the curve dips below the double top plate.

Next, we slide short lengths of backing along the rafters to determine the size and the location of the actual backing boards. Then we snap lines on the underside of the rafters to guide the installation. After assembling the backing boards on the floor, we attach them to the underside of the rafters with ring-shank nails or screws.

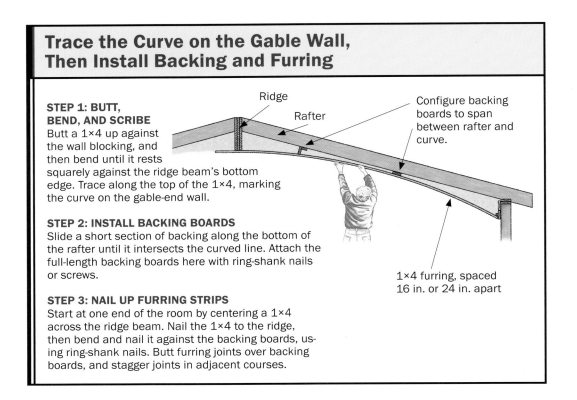

Trace the Curve on the Gable Wall, Then Install Backing and Furring

Ridge

Rafter

Configure backing boards to span between rafter and curve.

STEP 1: BUTT, BEND, AND SCRIBE
Butt a 1×4 up against the wall blocking, and then bend until it rests squarely against the ridge beam's bottom edge. Trace along the top of the 1×4, marking the curve on the gable-end wall.

STEP 2: INSTALL BACKING BOARDS
Slide a short section of backing along the bottom of the rafter until it intersects the curved line. Attach the full-length backing boards here with ring-shank nails or screws.

STEP 3: NAIL UP FURRING STRIPS
Start at one end of the room by centering a 1×4 across the ridge beam. Nail the 1×4 to the ridge, then bend and nail it against the backing boards, using ring-shank nails. Butt furring joints over backing boards, and stagger joints in adjacent courses.

1×4 furring, spaced 16 in. or 24 in. apart

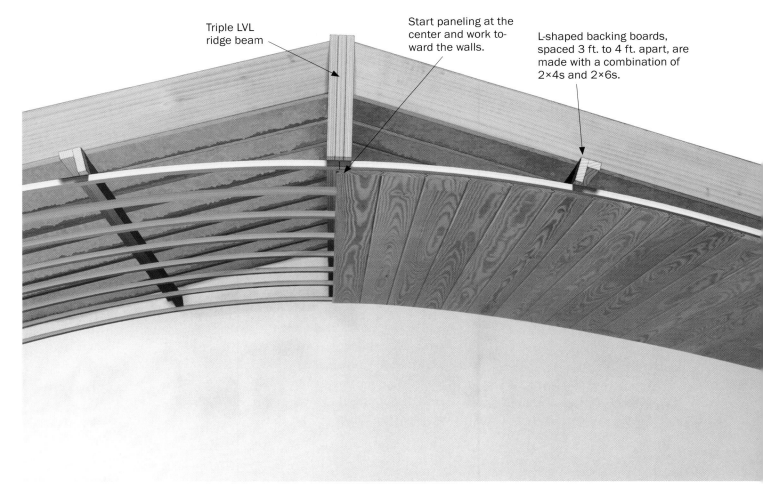

Triple LVL ridge beam

Start paneling at the center and work toward the walls.

L-shaped backing boards, spaced 3 ft. to 4 ft. apart, are made with a combination of 2×4s and 2×6s.

Offset the Splices in the Strapping

With the 1×4 furring strips, make sure to offset the splices in adjacent courses so that the overall curve of the ceiling can stay as fair as possible. Generally, the fairest curve should be across the center of the ceiling span, so it's smart to start with the clearest stock centered across every other rafter. This leaves short sections at the walls to fill later. Next, we install long pieces starting at the walls, spliced on the ridge, between the first set of curves. We cut and fill the short end pieces with knottier stock.

Using ring-shank nails (or even deck screws) to fasten the furring strips is worthwhile because the nails will be loaded in withdrawal from fighting the tension of all those tortured 1×4s. We often finish the ceilings with wood paneling, but two layers of ⅜-in. drywall work, too. When you're using tongue-and-groove paneling, spray-foam insulation ensures an effective air barrier above the paneling. A well-detailed drywall ceiling should stop air movement enough to allow you to use fiberglass or cellulose insulation.

Michael Chandler owns Chandler Design-Build (www.chandlerdesignbuild.com) near Chapel Hill, North Carolina.

Wood paneling adds detail. You can finish the ceiling with wood paneling or drywall.

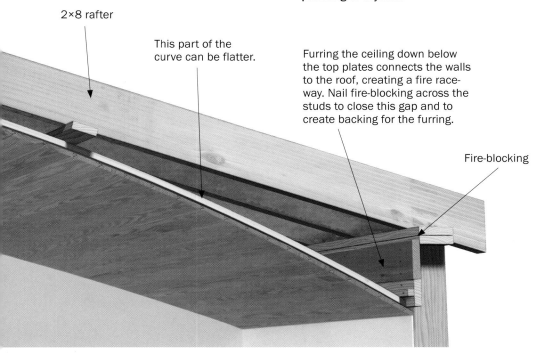

2×8 rafter

This part of the curve can be flatter.

Furring the ceiling down below the top plates connects the walls to the roof, creating a fire race-way. Nail fire-blocking across the studs to close this gap and to create backing for the furring.

Fire-blocking

Framing Cathedral Ceilings

■ BY BRIAN SALUK

I started framing houses years ago, before cathedral ceilings came into fashion. When asked to frame my first cathedral ceiling, I went at it much as I did any other roof. After bracing the walls plumb with left-over 10-ft. 2×4s, my crew set the ridge and the rafters, seemingly without a hitch. It was just another gable roof, only without the ceiling joists. When we finished setting the rafters, it was lunchtime.

I remember biting into my ham-and-cheese sandwich and looking back at the roof. I expected that feeling of satisfaction one gets looking on the results of a good morning's labor. Instead, I got a sinking feeling: The center of the ridge was sagging.

Back in the house, I saw that the weight of the roof pushing out on the walls had actually pulled some of the braces from the floor. Fortunately, it took only a couple of hours to jack the ridge level and pull the walls straight with a come-along. But I was lucky that the only serious loss that day was my uneaten lunch.

Brace the Walls to Resist Roof Thrust

The strength of a typical roof derives from the triangular shape made by the rafters and ceiling joists (see the top photo on p. 246). The ceiling joists tie the exterior walls together, resisting the outward thrust on the exterior walls. Because the joists tie the rafters together as a unit, the rafters carry the downward load on the ridge to the eave walls. Remove the joists, as with a cathedral ceiling, and two things happen. The rafters push out and bow the eave-wall plates, and the ridge becomes load bearing and sags because it isn't sized to bear a load (see the top photo p. 246).

Building cathedral ceilings means finding ways to duplicate the joist's function or eliminating the need for it, both during construction and as part of the permanent structure. Simply put, if you keep the bottoms of the rafters from spreading apart or if you keep the ridge from sagging, the roof will be strong and stable.

Simply put, if you keep the bottoms of the rafters from spreading apart or if you keep the ridge from sagging, the roof will be strong and stable.

Cathedral ceilings are tricky. The challenge boils down to keeping the walls from spreading during and after construction.

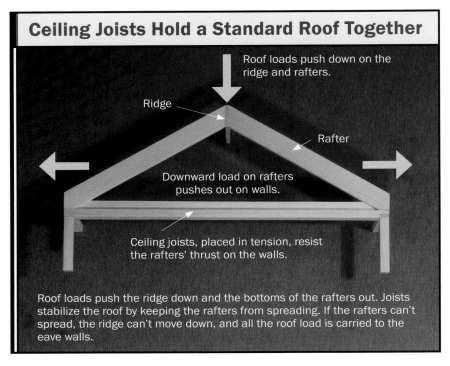

Ceiling Joists Hold a Standard Roof Together

Roof loads push down on the ridge and rafters.

Ridge

Rafter

Downward load on rafters pushes out on walls.

Ceiling joists, placed in tension, resist the rafters' thrust on the walls.

Roof loads push the ridge down and the bottoms of the rafters out. Joists stabilize the roof by keeping the rafters from spreading. If the rafters can't spread, the ridge can't move down, and all the roof load is carried to the eave walls.

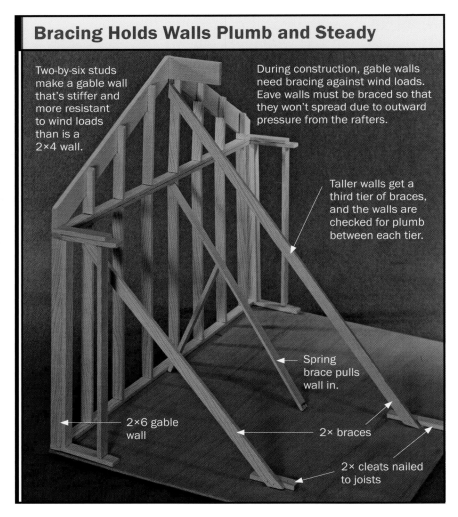

Bracing Holds Walls Plumb and Steady

Two-by-six studs make a gable wall that's stiffer and more resistant to wind loads than is a 2×4 wall.

During construction, gable walls need bracing against wind loads. Eave walls must be braced so that they won't spread due to outward pressure from the rafters.

Taller walls get a third tier of braces, and the walls are checked for plumb between each tier.

Spring brace pulls wall in.

2×6 gable wall

2× braces

2× cleats nailed to joists

Proper bracing is the most important consideration during construction. If the walls aren't properly braced, the rafters' thrust will bow the plates and their weight will sag the ridge. And the wind, blowing against the tall gable-end walls typical of cathedral ceilings, can knock down the whole assembly.

I brace the eave and gable walls plumb at least every 8 ft. with 2×4s half again as long as the wall is high, or 12 ft. for an 8-ft. high wall. One-story gable walls call for two tiers of braces, one high and one low (see the bottom photo at left). Two-story gable walls get three tiers of braces, and I check the walls for plumb between each tier. I sometimes use 2×6s for the longest braces. The braces are nailed to the tops of studs and to 2-ft.-long 2×4 cleats that are nailed to a floor joist. I use at least two 12d nails at each end of a brace.

Straightening the top plates usually involves pulling the walls in as well as pushing them out. I pull them in with spring braces made from long 2×4s nailed on the flat to the top of the top plates and to the floor at a joist. Jamming a shorter 2×4 under the middle of a spring brace bends it and pulls in on the wall.

T-Shaped Posts Support the Ridge

If the ridge of a cathedral ceiling can be kept from sagging, the rafters can't push out the wall plates and the roof stays put. This is the principle behind the structural-ridge roofs that I will be describing later. With the other types of cathedral ceilings, though, the ridge board isn't load bearing except during the construction process. Because of this, the ridge board is not sized to take a load and can sag from the weight of the rafters during construction.

To avoid this sagging, I support the ridge with T-shaped posts made by nailing a 2×4 on edge to the center of a 2×6. The T-shape of the post resists buckling under load better than does a single piece of lumber. I space the posts no more than 12 ft. apart and make sure that each one sits over well-supported floor joists. If I have doubts about a post sitting on one joist, I stand the post on a 2×10 or 2×12 laid flat over several joists to spread the load.

Rafters Must Make Room for Insulation

Many framers lay out roofs so that opposing rafters are staggered, making it easy to nail through the ridge into rafter ends. But it's usually best to align opposing rafters in a cathedral ceiling to allow subsequent members to be nailed evenly, instead of at an angle. I keep this in mind when I lay out the mudsills so that floor joists and studs stack under the rafters.

Once I start framing and the customers can finally begin to see the house three-dimensionally, it's common for them to ask if a flat ceiling could become cathedral. If the rafters and ceiling joists aren't already cut, accommodating this request is usually a simple matter of stepping up the original rafter size at least one dimension. For example, I can use 2×8 rafters on a 28-ft. wide house that has a conventional roof. If the roof were changed to a cathedral style, I'd use at least 2×10 rafters. The reason is twofold. Extra heft helps to keep the rafters from sagging over time. And without flat ceiling joists, the insulation goes in the roof. The rafters must be wide enough to accommodate the insulation plus space for ventilation.

I have to be selective with rafter material when building cathedral ceilings. The underside of the rafters forms the ceiling plane, so any rafter material with extreme crowns that might show through the finish ceiling gets culled.

The bird's mouths in cathedral rafters have to be cut so that the bottom of the rafter intersects the corner of the top plate. If it doesn't, there would be an area between the ceiling plane and the wall with no nailing for drywall.

Working from Scaffolds Speeds Construction

I find it's faster and safer to build cathedral ceilings from scaffolds than to work from ladders. I site-build scaffolds from framing lumber and plywood. Ideally, the scaffold should be high enough so that you can nail the rafters to the ridge and low enough to ease nailing the joists, if any, to the rafters.

The main supports for the scaffolds I build are goalpost-shaped assemblies with 2×4 legs and at least 2×6 horizontal members. I cross-brace these with more 2×4s and space them about 8 ft. apart. Two-by-tens are laid across the goalposts and covered with sheets of plywood. The posts should be high enough to support toeboards and guardrails.

Hip and Valley Rafters Can't Hang below Commons and Jacks

Hip and valley rafters are often sized deeper than the rest of the rafters because they carry the combined loads of the jack rafters. Normally, nobody cares if a beefed-up hip or valley rafter hangs down below the other rafters into the attic. But with a cathedral ceiling, a deep rafter would protrude through the finish ceiling. Because of this, if your plans call for an oversize hip or valley rafter, they may have to be made from two smaller members nailed together. If you have any doubts about hip-rafter or valley-rafter size, consulting an engineer is wise. Alternatively, you could fur the ceiling out to the level of the protruding hip or valley. Finally, ventilation along a hip or valley requires some thought (see the photo on p. 248).

Once all the rafters are up, I usually sheathe the roof. The ridge is still supported with temporary posts, so the roof assembly is strong enough for my men to work on.

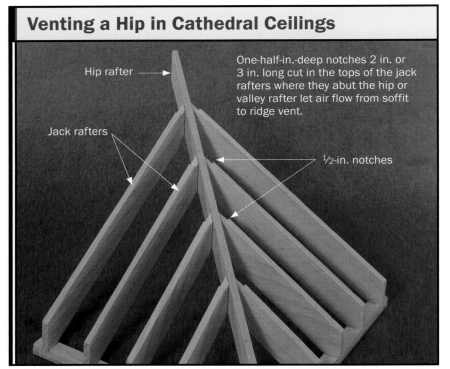

Venting a Hip in Cathedral Ceilings

Hip rafter

Jack rafters

One-half-in.-deep notches 2 in. or 3 in. long cut in the tops of the jack rafters where they abut the hip or valley rafter let air flow from soffit to ridge vent.

½-in. notches

Sheathing the roof at this point stiffens it and takes the bounce out the rafters, making it easier to nail the subsequent members to them.

Raising the Ceiling Joists Is the Simplest Cathedral Ceiling

Raising the stable triangle of joists and rafters upward is not much more complex than framing a standard gable roof (see the top photo on the facing page). It's probably the least expensive route, and the mix of angles and flats makes for an interesting ceiling. But if the triangle becomes too small, it can't stabilize the roof. I'll raise these joists about one-third of the distance from the top of the wall to the underside of the ridge. Lower is stronger.

Ceiling joists can often be raised higher than this, but a variety of factors comes into play. The room width, the roof pitch, and the snow load all must be considered. It's wise to consult a structural engineer before raising the joists higher.

I frame this roof much as I would a normal gable roof, starting with the end rafters, the gable walls, and the ridge. After supporting the ridge with a T-post, my crew sets the rafters.

After deciding their height, I install the joists. They must be level and in plane with each other. I measure up from the floor and mark the height on both gable walls. A joist is nailed at both ends of the room and checked for level.

I locate the rest of the joists with strings, rather than by snapping chalklines on the underside of the rafters. The rafters are never crowned exactly the same; thus, a chalkline won't be straight, and the ceiling won't be flat. I cut blocks from a piece of scrap and nail them atop the ends of the gable joists. I string a line on each side of the room from these blocks and space the remaining joists down from the lines with other blocks (see the bottom photo on the facing page). The joists don't touch the string, reducing the chance of accidentally pushing it out of line. The strings are set above the joists so that my crew doesn't have to wrestle them over the strings. Variation in joist width isn't usually a problem, particularly if all the stock comes from the same pile of lumber. The joists are nailed to the rafters with at least six 12d nails in each joint. I cut the joists to the roof angle so that there is more wood to nail into than if the joists were square-cut. I cut them just short enough so that they won't touch the roof sheathing. This way, the rafters won't shrink past the joist ends, creating bumps in the roof.

If the span is sizable, I use wider joists. For spans up to 14 ft., 2×6s are fine; beyond that, I increase to 2×8s. If the joists span more than 12 ft., I nail a 2×4 flat to the top of the joists, running perpendicular to the joists and centered in the span. A 2×6 on edge is nailed to the 2×4, creating a strongback. I place the strongback material on top of the joists before installing all of them. Otherwise, I won't be able to get the material up there at all.

A variation on this ceiling is to double the joists on every third rafter pair and leave

Raised Ceiling Joists Are the Simplest Cathedral Ceiling

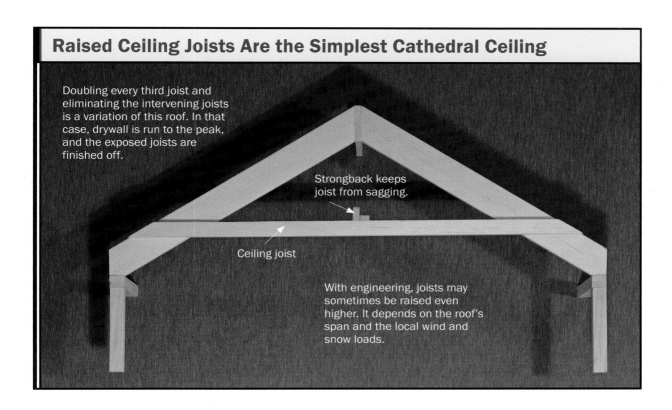

Doubling every third joist and eliminating the intervening joists is a variation of this roof. In that case, drywall is run to the peak, and the exposed joists are finished off.

Strongback keeps joist from sagging.

Ceiling joist

With engineering, joists may sometimes be raised even higher. It depends on the roof's span and the local wind and snow loads.

Strings Help the Carpenters to Align the Ceiling Joists

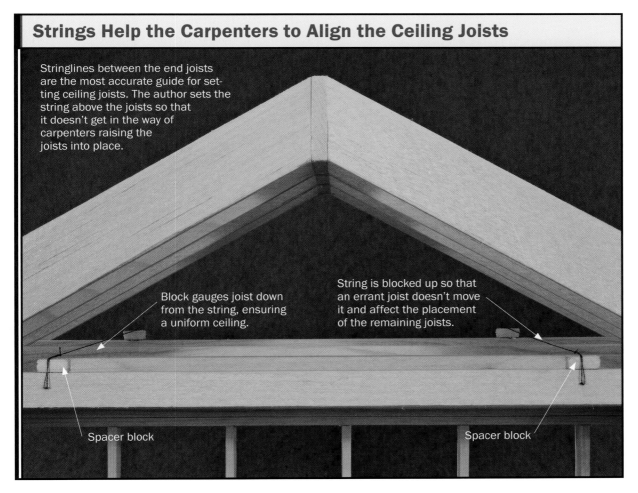

Stringlines between the end joists are the most accurate guide for setting ceiling joists. The author sets the string above the joists so that it doesn't get in the way of carpenters raising the joists into place.

Block gauges joist down from the string, ensuring a uniform ceiling.

String is blocked up so that an errant joist doesn't move it and affect the placement of the remaining joists.

Spacer block

Spacer block

out the intervening joists. Similar caveats about not raising the joists more than one-third the roof height apply. On this ceiling, the drywall goes all the way to the peak. The doubled joists are exposed, and either drywalled or finished with trim stock.

Scissors Trusses Can Be Site-Built

A scissors truss consists of two opposing rafters braced by two pitched ceiling joists (or truss chords) that resemble lower-slope rafters (see the photo below). The chords cross at the ceiling's peak and continue upward to lap the rafters. This ceiling works well when the customer wants an unbroken ceiling plane right up to the peak. It's also good if the client wants the ceiling to be a shallower pitch than the roof is.

The chord's pitch shouldn't exceed two-thirds of the rafter's pitch. In other words, if the rafters are a 9-in-12 pitch, the chords should be a 6-in-12 or lower pitch. The steeper the pitch of the chords, the less effective they are at bracing the rafters. I make the chords one size smaller in depth than the rafters.

Framing a scissors-truss roof begins similarly to framing a raised-joist roof. Set the gables and the ridge. Brace the ridge, set the rafters, and partially sheathe the roof. Here, it's especially important to lay out the rafters so that they align at the ridge.

The gable rafters are supported by walls, so there is no need to brace them with chords. The gable-end chords essentially

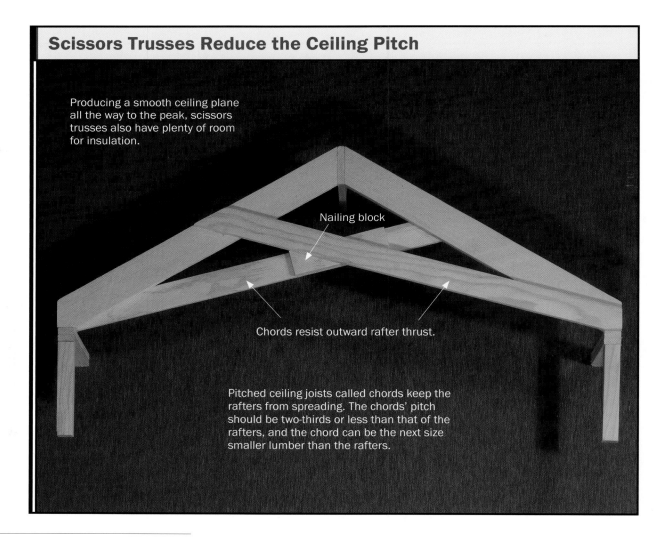

Scissors Trusses Reduce the Ceiling Pitch

Producing a smooth ceiling plane all the way to the peak, scissors trusses also have plenty of room for insulation.

Nailing block

Chords resist outward rafter thrust.

Pitched ceiling joists called chords keep the rafters from spreading. The chords' pitch should be two-thirds or less than that of the rafters, and the chord can be the next size smaller lumber than the rafters.

serve as drywall nailers and are nailed to the gable walls. I lay them out just like common rafters, without deducting for a ridge. After nailing up the gable-end chords, I cut the bird's mouth on a piece of chord stock that's long enough to span from the wall to the opposing rafter. I hold this chord stock in place, even with one of the gable-end chords. By marking the chord stock where it laps the opposing rafter, I have the pattern for the rest of the chords.

To line up the chords, I string two lines from the top of the end chords, just as I did with the raised-ceiling-joist roof. The chords are nailed on opposing sides of rafter pairs with six 12d nails per joint. I also toenail them to the wall plate. The chords are lined up on the strings and nailed to the rafter on the far side of the ridge. Where the chords cross, they're the thickness of the rafter apart. I nail a 2-ft. block of the chord material flush with the bottom of one chord and nail the second chord to the block.

Design the Ridge as a Beam, and No Joists Are Needed

Another approach to cathedral ceilings is to make the ridge a beam that's stiff enough not to sag under load (see the photo at right). I build this type of roof when a ceiling that climbs cleanly to the peak at the roof pitch is wanted.

A structural ridge creates point loads that must be carried through the gable wall to the foundation with continuous, stacking framing. Headers in this load path need to be sized accordingly, and their studs may need beefing up, too.

I balloon-frame particularly tall gable walls. The studs in balloon-framed gable walls reach from the bottom plate on the first floor to the top plate just below the rafters. Balloon-framing avoids the plates at the various floor levels common to platform-framed walls. Plates can act as a hinge, weakening tall walls. To stop the chimney effect these continuous

Structural Ridge Beam Keeps Roof from Sagging

Because the ridge carries roof loads to a gable wall, gable-wall framing under the ridge must establish an uninterrupted load path. Headers in this path should be engineered. Use LVL headers, and the manufacturer will usually engineer them for free.

½-in. gap for ventilation

Structural ridge doesn't sag under design load, so rafters don't push out on walls.

Three-member structural ridge, often composed of engineered lumber

Continuous 2×6 framing carries load to foundation.

stud cavities can have in a fire, codes specify fire blocking at least every 8 ft. and where the wall intersects floors and ceilings.

In areas where gable walls are subject to high wind loads, I frame gable walls with continuous 2×6 LSL studs. LSL is factory made by shredding lumber and gluing the strands back together. LSL is denser and stiffer than solid-sawn lumber, and makes for a stronger but more expensive wall.

I avoid large, single-member ridge beams. They're heavy and often must be placed with a crane. I prefer to assemble in place two, three, or even four full-length LVL members that can be lifted by hand. LVLs can span greater distances than standard lumber and are made from material similar to LSL studs. And LVL manufacturers will usually size the beam for you at no extra cost.

Individual ridge members longer than 24 ft. are usually too heavy to lift by hand. In that case, I'll assemble the beam on the ground and lift it with a crane. It's important to build beams straight; once nailed, they're nearly impossible to straighten.

Before nailing together a multimember beam in the air, I set the gable rafters and wall. On 2×6 gable walls, my crew sets the first beam member between the gable rafters just like a ridge board. The subsequent members are cut shorter so that they butt to the inside of the gable rafters. Even after deducting 1½ in. for the gable rafter, the beam has a full 4 in. of bearing. I stagger these shorter members down so that they're about ½ in. lower than the top of the rafters. This method allows air to flow to the ridge vent. After nailing the beam together, I measure, cut, and then install the post under the beam to carry the load downward.

For 2×4 gable walls, all members of the ridge beam must run through the entire width of the wall to gain sufficient bearing. This means that all the beam members have to be placed at the same time as the gable rafters, a trickier operation. Because of this situation and because 2×6 gable walls are stiffer, I rarely build 2×4 gable walls when using a structural ridge.

Show a Finished Beam beneath the Ridge

This roof goes up similarly to the previous example, except that the beam is installed below and supports a standard ridge (see the photo on the facing page). I build this type of roof when the customer wants to show a large finished beam or when the ridge beam is so deep that it would hang below the rafters anyway. In that case, I often put collar ties just below the beam for drywall nailers. This eliminates the need to drywall and finish that awkward triangular space between the rafters and the side of the beam.

Shorter beams that are light enough to be handled by a couple of carpenters can be installed after the rafters are set. With longer beams, however, especially big single-member beams, it's easier to set the beam first, then build the roof around it.

Again, the first step is building the gable walls and setting the gable rafters. The wall must have a post to support the beam, just as in the structural-ridge type of roof. I cut the gable rafters normally and set the beam within them by hand or by crane.

If this beam is to show, I treat it with care. I hoist it with nylon slings instead of chains, which can mar the surface. And I don't nail temporary braces to the finish face. The nail holes might show, and worse, if the nails rust, they'll deeply stain the beam.

The rafter tops will have bird's mouth cuts in them that fit over the beam. I don't toenail through the upper seat cut; this usually splits the top of the rafter. Rather, I nail the rafter to the ridge and toenail the ridge to the beam. When laying out this seat cut, I allow for the height of the ridge board plus ¼ in. or so. The rafters don't have to touch the beam because the ridge does. This ¼ in. allows a bit of play that simplifies setting the rafters.

Brian Saluk is a custom homebuilder from Berlin, Connecticut.

Beam under Ridge Carries Roof Load

A variation of a structural ridge, an exposed beam supports the ridge and carries roof loads to the gable walls. Often finish-grade material, this beam calls for careful handling.

Bird's mouths

Exposed beam

Continuous 2×6 framing carries load to foundation.

Ceiling Remodel: From Flat to Cathedral

■ BY MIKE GUERTIN

Most of the homes in my neighborhood are ranch style, built in the late 1950s and early 1960s. They all seem to have low ceilings that often measure 88 in. high, which makes them feel cramped and dark. Such a low ceiling is especially out of scale in a 16-ft. by 18-ft. family/living room, as was the case in a ranch I was remodeling. A quick peek in the attic confirmed that the framing was conventional rafters, not trusses. This meant that I could transform the room by adding a structural ridge and reframing the ceiling. The new ceiling would add about 42 in. of height at the center and improve the character of the room. By recycling the existing ceiling joists, I'd need to buy only drywall and a bit of lumber.

Finish the Design before Starting Work on the Demolition

I had two options for the ceiling design (see the illustrations on pp. 256–257): a monoslope vault that ran uninterrupted from the exterior wall to the interior bearing wall; or what I would call a gable vault, which created a false ridge in the middle of the room. This second option seemed in keeping with a ranch house, more so than the monoslope of 1980s contemporary-style homes. To gain the greatest height in the new ceiling, I decided to insulate and drywall right to the underside of the existing 2×6 roof rafters.

Recycled ceiling joists make good rafters. Careful demolition saved me a trip to the lumberyard. After finding the centerline, I used the old joists as the new rafters on the gable vault. The open gable end was framed last.

Because I removed the ceiling joists that reinforce the walls against the weight and thrust of the roof, I added a structural ridge. The double 2×12 beam spans from the outer gable wall to an interior bearing wall and is supported at both ends by king and jack studs. I also added new rafters to frame the other half of the gabled ceiling.

The new structural ridge beam, a pair of 2×12s, spans from the outer gable wall to the middle of the bearing wall and is supported at both ends by king and jack studs.

Pocket framed into gable wall

CATHEDRAL OPTIONS

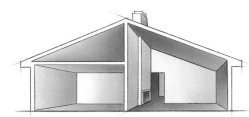

A **monoslope** ceiling extends from the exterior wall's top plate to the roof ridge.

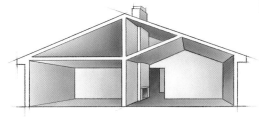

A **gabled** ceiling creates a false ridge at the room's centerline.

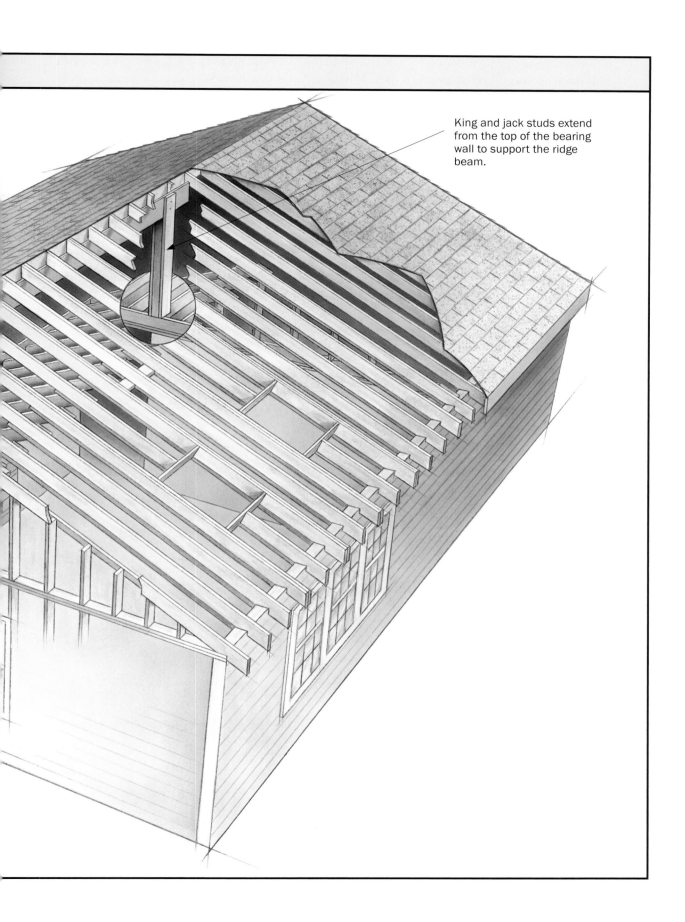

King and jack studs extend from the top of the bearing wall to support the ridge beam.

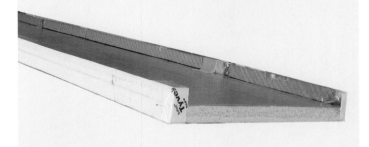

Combine insulation and venting. Custom vent chutes made from 1-in. rigid-foam insulation are fit in each rafter bay; 1-in.-wide strips taped along the edges space the panels off the roof sheathing and form the vent channel. Added later, 6 in. of blown cellulose will create a ceiling insulated to R-25.

Before tearing out the plaster ceiling, I moved the rock-wool ceiling insulation to another part of the attic to minimize demolition mess. I also rerouted electrical wiring that crossed the ceiling joists. In the attic, I slipped in a pair of 2×12s long enough to span from the gable end to the bearing wall in the hall that would support the roof once the ceiling joists and strapping were removed (see the sidebar on the facing page). The old ceiling joists act like rafter ties for the roof, resisting the outward thrust of the rafters and preventing the exterior walls from bowing outward. A new structural ridge beam beneath the ridge would accomplish the same goal.

Make a Better False Ridge with Straight Lines

I used a plumb bob at both ends of the room to locate the center **1**. then snapped a chalkline on the ceiling between them **2**. The chalkline indicated where the bottom of the false rafters would cross the existing ones to form the vault. I drove screws into the rafters partially to support the rafter tops (see the inset photo on p. 255), and I toenailed the tails into the double 2×6 that supported the old ceiling along the face of the interior fireplace wall **3**. Each rafter top then was spiked to the primary rafters with four 12d nails.

Careful Demolition Saves Both the Floor and Lumberyard Expenses

I covered the hardwood floor with moving blankets, then a layer of 6-mil plastic for protection and for easier cleanup. The old 16-in.-wide gypsum board and ½-in.-thick plaster are easiest to remove by pulling them down along the seams. I usually start by knocking a small hole with a wrecking bar and letting the weight of the plaster assist the removal, working my way across the ceiling. Don't forget to wear a good respirator and eye protection when doing this kind of work.

The strapping usually can be pried off the ceiling joists without much of an effort. Rather than beating everything to a pulp with a wrecking bar after I'd cleared the rafters, I used a reciprocating saw fitted with a bimetal blade to cut the toenails pinning the joists to the plates; this process prevented the joist ends from splitting. Once the nails had been removed, the strapping and the joists would provide much of the lumber needed to frame the new ceiling.

Protect the floor during demolition. Hardwood floors are protected by moving blankets under a layer of 6-mil plastic, which also makes cleanup easier. The old 16-in.-wide gypsum board and ½-in. plaster are easiest to remove by pulling down along the panel edges. Pieces usually break into 16-in. by 8-ft. slabs. Careful demolition also preserves strapping and rafters for later use.

Retrofit a Structural Ridge Beam before Demolition

Before tearing out the ceiling, I brought a double 2×12 beam to the attic to support the portion of the ridge where I would remove the ceiling. The existing ridge board bears directly on the top edge of the new beam. With the beam's far end set in a pocket at the gable, I raised the inboard end of the beam into place by using a pair of 2×6 king studs as a lifting jig. These uprights had holes drilled every 6 in. for a ½-in.-dia. steel rod I inserted as a safety stop **1**. At each stop, I slid the rod through the holes to support the beam. After an initial manual lift by hand, I used a hydraulic jack to raise the beam level **2**. Finally, tight-fitting jack studs were driven under the beam and screwed to the king studs **3**.

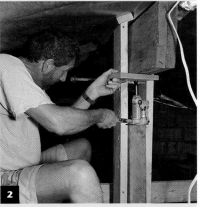

Frame Now for Skylights Later

Sister rafters without the tails are cut from the old ceiling joists. After the roof nails protruding through the sheathing are bent flat adjacent to the rafters, the top of a sister rafter is aligned alongside the existing one. Oriented this way, the only friction to overcome is at the seat cut and the top cut. Clamps draw the rafter tops together, and I rap the bottoms with a hammer until they're tight **1**. Nails are staggered and spaced 12 in. apart to hold each pair together **2**.

Make the original roof better. Older roofs like this one were underinsulated and underbuilt, so this was a good time to rectify the situation. I also planned to install skylights at some point in the future, so this was my opportunity to add the extra framing.

Shims Correct Roof Sag, and New Rafters Extend to a False Ridge

The plans called for two skylights to be installed after the roofing was replaced, so this was the perfect time to frame the roof in preparation. I doubled the rafters on each side, then added the headers above and below the future skylight locations.

The roof rafters had an obvious sag in the middle of the room. Rather than ignoring it, I ripped tapers on some 2×3s and nailed them to the underside of each rafter to create a uniform ceiling plane. I recut the old ceiling joists and used them to frame the other side of the new ceiling at the same 5-in-12 pitch of the existing roof. I sidenailed the false rafters to the roof rafters at the top and toenailed them to a beam at the bottom. The end wall was framed with 2×4 studs and cleated along the top to carry the ceiling drywall.

After installing new lighting and air-conditioning registers, I hung and finished the drywall (see the top photo on the facing page), blew in cellulose insulation, then painted. The results (see the bottom photo on the facing page) were better than I expected.

Mike Guertin is a builder, a remodeling contractor, and a contributing editor for Fine Homebuilding *from East Greenwich, Rhode Island. His website is www. mikeguertin.com.*

Straighten a Sagging Roof with Tapered Shims

The roof above the living room sagged to the point where I couldn't ignore it. The ridge was bowed, and the rafters had an obvious sag in the middle of the room. To determine the severity of the problem, I set two stringlines as guides, one just above the top wall plate and one at the centerline of the room. Both lines were blocked out 2½ in. from the face of the rafters and strung from opposite sides of the room. After consecutively lettering the rafters, I clamped a long, straight 2×3 to the side of each one. I lettered each 2×3 with the same letter as the rafter and aligned it with the stringlines at each end. After using the edge of the rafter to scribe the shim, I unclamped the 2×3s, ripped them along the scribe lines, and screwed each one to the underside of its rafter. In addition to straightening the interior plane of the roof, the 2×3s added extra depth to the rafter bays for extra insulation. Alternatively, I could have sistered each 2×6 rafter with a 2×8 or 2×10, but that would have used more lumber and also would have reduced the width of the bays and their insulation.

Blocks hold the string at 2½ in. from the rafters.

Rafters, 16 in. o.c.

A new 2×3 is clamped to each rafter, oriented to the stringlines, scribed, ripped, and then fastened to the rafter face.

Stringline at ceiling's midpoint

Stringline at plate

Rent a helper. Drywall lifts are usually available at local rental outlets and are worth every penny. They save time and make solo installation of even a 12-ft.-long sheet an easy task.

With the addition of skylights, the finished ceiling adds light and space to the living room.

Do You Need an Engineer?

When I contemplate rearranging the major structure of a house, it means that I'll soon be talking permits with the local building department. In this case, I wanted to remove the ceiling joists and rafter ties. I employed the option described in R802.3.1 (IRC 2003): "Where ceiling joists or rafters are not provided at the top plate, the ridge formed by these rafters shall also be supported by a girder designed in accordance with accepted engineering practice." I discussed my plan with the building inspector, and he was satisfied that I was maintaining the integrity of the structure I was working on.

Depending on your location and the scope of the job, you also might need to call in a structural engineer. Some building departments require it, and it's sometimes just better to make sure modifications are safe.

Open Up the Ceiling with a Steel Sandwich

■ BY MICHAEL CHANDLER

My company specializes in designing and building small homes. I like to vault ceilings in small rooms—bedrooms and screened porches, commonly—that normally would have flat ceilings.

People love cathedral ceilings because they add drama and presence to interior spaces. A vaulted ceiling can make a small room look large rather than cramped and confined. So why are cathedral ceilings usually reserved for living rooms or other large public areas, whereas small rooms get stuck with flat ceilings? Cost.

Scissors trusses or structural ridge beams are the common, expensive methods for incorporating cathedral ceilings in houses. For the past few years, though, my company has overcome this economic problem by using angled steel flitch plates supplied by a steel fabricator. As shown in the illustration on p. 266, the V-shaped plate is angled to match the roof peak, and it's sandwiched between common rafters. Angled steel flitch plates work well for pyramid roofs but can be used to open gable roofs, too.

Scissors Trusses Don't Cut It

For years, I've been building vaulted ceilings using scissors-truss kits from my local truss supplier, but scissors trusses have a number of drawbacks: less space for insulation at the top plate, scheduling headaches, and cost. Did I mention that they're also a pain in the neck to install? Getting the little trusses to line up for a smooth roof and ceiling is fussy at best; my crew and I often have to strap the ceilings with 1×4s to get them smooth enough for drywall. The truss package for the last pyramid hip ceiling I built, for a 16-ft. by 16-ft. bedroom, cost $1,600*. The three-week lead time added insult to injury. There had to be a better way.

My engineer came up with an angled flitch-plate detail that creates a V-shaped beam capable of supporting 2×12 framing and that adds only $180 to the framing costs. It's like a 160-lb. Simpson Strong Tie. The welding must be done by an American Welding Society certified welder using a

Bolted between rafters, an angled steel flitch plate can eliminate the need for rafter ties in cathedral ceilings.

Engineering a Cathedral Ceiling

The weight of a roof naturally causes rafters to thrust outward, which can push the walls apart. In a typical house ceiling, joists prevent the walls from spreading. In a cathedral ceiling, there are a few common ways to offset this outward thrust.

A **structural ridge** is a beam large enough to span the length of the house and support the weight of the roof.

Rafter ties provide tension force, which offsets the outward thrust and works for hip or gable roofs. The downside: horizontal framing members.

Scissors trusses are engineered to distribute forces internally on hip or gable roofs. The downsides: reduced ceiling pitch, difficult to drywall on hip roofs, expensive.

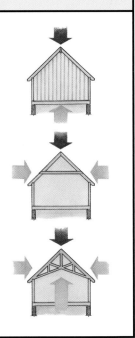

A Welded Steel Plate Shoulders the Roof Load and Unclutters the Ceiling

A typical flitch plate is a straight piece of steel bolted between framing lumber. This modified design for a roof beam supports the framing while holding the walls together. This design works on hip roofs, gable roofs, or even shed-dormer retrofits. The benefits: inexpensive, open cathedral ceiling, readily available.

Bolt flitch plate between common rafters.

$^{11}/_{16}$-in. hole

$^3/_8$-in. steel flitch plate with holes cut by fabricator

Full-penetration weld must be done by a certified welder.

$^5/_8$-in. by 4-in. bolt

Fill in below steel with $^3/_8$-in. plywood.

2×10 rafter

---

full-penetration weld; you can't just do it on site. The first time I brought one of these angled-flitch drawings to my welder, he wanted a plywood template. Now I can fax him a diagram and he has the flitch plate ready for me to pick up in a week, complete with clean holes.

This flitch solves all the problems of scissors trusses: It has a full 10 in. for insulation over the top plate, it keeps the ceiling the same pitch as the roof, and it ensures that the roof and ceiling planes are smooth every time. Yes, the steel is heavy, and getting all those holes to line up takes some planning. But the assembled flitch beam is much less awkward to handle than scissors trusses, and it costs about $1,000 less for that 16-ft. by 16-ft. room.

Get help. Installing scissors trusses requires one or two helpers.

The Holes Are Bigger Than the Bolts

The holes in the steel should be cut by the fabricator, preferably with a high-pressure hydraulic "piranha" punch. Do not blast holes in the steel with a cutting torch because the holes will be ragged and will catch at the drill bit while you're using the steel as a template to drill holes into the wood.

When drilling through the rafters, use the same-size drill bit as the hole in the steel. This step ensures good contact for the bolt between the wood and the steel. The bolt, however, should be slightly smaller than the hole. An $1\frac{1}{16}$-in. hole in the steel requires a $\frac{5}{8}$-in. bolt. Undersizing the bolt more than that weakens the connection because the steel can move in the sandwich; undersizing the holes in the rafter can cause the wood to split when the bolts are tightened.

The welding must be done by an American Welding Society certified welder using a full-penetration weld; you can't just do it on site.

Tack the rafters flat. After laying the rafters atop the wall plates and a temporary catwalk, nail the plumb cuts tightly together. Align the seat cuts to their layout marks on the wall plates, and toenail.

The steel is shorter than the rafters. Fill in the ends of the steel plate with $\frac{3}{8}$-in. plywood and 1-in. roofing nails. After the top rafter is bolted in place, 8d nails driven from each side will hold the tails together.

Use the steel as a drilling template. After laying the steel in place on top of the rafters, bore a hole at each end of the steel, then insert a bolt to hold the steel in place while the remaining holes are drilled.

Complete the sandwich. Clamp the top rafters in place and drill up through the previously drilled holes. Again, make sure the plumb cuts are tight before drilling the holes.

Assemble the Beam in Place

Because the flitch weighs a lot (160 lb., in this case), the installation is not a one-person job. You'll need a helper or two to build this beam in place. I set up a temporary catwalk down the center of the room with a support post underneath. The catwalk needs to be stiff enough to support the framing lumber, the steel, a helper, and me.

With the catwalk ready, I set up my sandwich shop. I lay a couple of common rafters flat and tack them in place on the catwalk and wall tops (see photo 1 on p. 267). I tack the rafters to the wall plates so that the nail will act as a hinge—that is, the nail will hold the rafter in place during assembly and keep it from sliding when it's time to stand up the beam.

Next, I lay the steel on the rafters and drill down. To prevent the plate from shifting around, I drop several bolts into holes as I go (see photo 2 on p. 267). Having a helper comes in handy for more than just lifting the steel. A helper minimizes my number of trips up and down the ladder for bolts, nuts, washers, and plywood filler strips. Have a helper cut and fit ⅜-in. plywood spacers as you drill holes (see photo 3 on p. 267).

With all the holes drilled in the lower-rafter pair, I pull the bolts (except for one in

A Nail Holds the Beam while It's Tilted Up

Place the first rafter exactly on the layout line so that when it's rolled up the rafter will be in the right place. Drive a 16d nail at an angle through the rafter and into the top plate. As the beam is lifted, the nail bends but keeps the rafter on its layout line. Have 2×4 braces on hand to nail into the beam after it's upright.

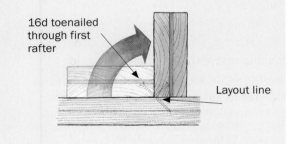

16d toenailed through first rafter

Layout line

the tail of each rafter), lay the second rafter pair in position, and clamp them in place. The bolts remaining keep the parts aligned while the first couple of holes are drilled in the second rafter pair. Drill up from below at the peak, and set bolts in (see photo 4 on p. 267). After removing the first pair of bolts, I snug the second pair down to prevent drill vibration from causing the parts to lose alignment.

Stand and Brace the Beam

Once all the bolts are tightened down, tilt the assembly upright and brace it well. It's not that heavy as long as the ends are well secured on the top plates. Have the bracing ready to nail as soon as the assembly is plumbed. Although 2×4 braces work fine, I like to set a perpendicular common rafter and a couple of hip rafters as bracing if I'm unable to finish the roof framing that day.

The rest is just regular hip-roof stick-framing except that the rafters are loaded in shear at the peak, so you need to use a hanger designed for shear load.

This type of roof assembly requires more uplift reinforcement than a comparable roof with ceiling joists that are used as collar ties, because there's less meat for toenailing into the top plate. For this reason, exceeding the code requirement for rafter tie-downs is a great idea.

Prices are from 2006.

Michael Chandler owns Chandler Design-Build (www.chandlerdesignbuild.com), near Chapel Hill, North Carolina.

Use steel framing connectors for shear resistance. The steel flitch-beam assembly shoulders the shear load in one direction, but the intersecting hip rafters should be supported with a special framing anchor (the Simpson HRC22 hip-rafter connector) as well.

CREDITS

p. iii: Photo by Mike Rogers, courtesy *Fine Homebuilding*, © The Taunton Press, Inc.; p. iv, John Fournier, courtesy *Fine Homebuilding*, © The Taunton Press, Inc.; p. v (left) Mike Rogers, courtesy *Fine Homebuilding*, © The Taunton Press, Inc.; (right) Roe A. Osborn, courtesy *Fine Homebuilding*, © The Taunton Press, Inc.; p. vi: (left) courtesy Woodsmart Solutions, Inc.; (right) Chris Green, courtesy *Fine Homebuilding*, © The Taunton Press, Inc.; p. 1: (left) Chris Green courtesy *Fine Homebuilding*, © The Taunton Press, Inc.; (right) Roe A. Osborn, courtesy *Fine Homebuilding*, © The Taunton Press, Inc.; p. 2: (left) Charles Bickford, courtesy *Fine Homebuilding*, © The Taunton Press, Inc.; (right) Seth Tice-Lewis, courtesy *Fine Homebuilding*, © The Taunton Press, Inc.; p. 3: Ron Ruscio, courtesy *Fine Homebuilding*, © The Taunton Press, Inc.

The articles in this book appeared in the following issues of *Fine Homebuilding*:

p. 6: 10 Rules for Framing by Larry Haun, issue 158. Drawings by Christopher Clapp, courtesy *Fine Homebuilding*, © The Taunton Press, Inc.

p. 16: Framing with a Crane by Jim Anderson, issue 140. Photos by Mike Rogers, courtesy *Fine Homebuilding*, © The Taunton Press, Inc.; except p. 17 photo by Andy Engel, courtesy *Fine Homebuilding*, © The Taunton Press, Inc.; and pp. 18 (bottom right), 19 (top), and 20 (top) photos by Jim Anderson, courtesy *Fine Homebuilding*, © The Taunton Press, Inc.

p. 24: All about Headers by Clayton DeKorne, issue 162. Photos by Roe A. Osborn, courtesy *Fine Homebuilding*, © The Taunton Press, Inc.; Drawings by Dan Thornton, courtesy *Fine Homebuilding*, © The Taunton Press, Inc.

p. 34: Fast and Accurate Framing Cuts without Lines by Larry Haun, issue 118. Photos by Roe A. Osborn, courtesy *Fine Homebuilding*, © The Taunton Press, Inc.

p. 40: Nailing Basics by Larry Haun, issue 110. Photos by Roe A. Osborn, courtesy *Fine Homebuilding*, © The Taunton Press, Inc.; Drawings by Jim Meehan, courtesy *Fine Homebuilding*, © The Taunton Press, Inc.

p. 50: Anchoring Wood to a Steel I-Beam by John Spier, issue 180. Photos by Krista S. Doerfler, courtesy *Fine Homebuilding*, © The Taunton Press, Inc.; Drawings by Bob La Pointe, courtesy *Fine Homebuilding*, © The Taunton Press, Inc.

p. 53: Common Engineering Problems in Frame Construction by David Utterback, issue 128. Photos courtesy the Western Wood Association; Drawings by Christopher Clapp, courtesy *Fine Homebuilding*, © The Taunton Press, Inc.

p. 62: The Future of Framing Is Here by Joseph Lstiburek, issue 174. Photos by Daniel S. Morrison, courtesy *Fine Homebuilding*, © The Taunton Press, Inc.; Drawings by Chuck Lockhart, courtesy *Fine Homebuilding*, © The Taunton Press, Inc.

p. 70: Engineered Lumber by Scott Gibson, issue 150. Photos by Scott Phillips, courtesy *Fine Homebuilding*, © The Taunton Press, Inc.; except p. 71 by Brian Vanden Brink, courtesy *Fine Homebuilding*, © The Taunton Press, Inc.; and p. 75 courtesy APA-The Engineered Wood Association.

p. 80: Reconsidering Rot-Resistant Framing Material by Scott Gibson, issue 186. Photos by Krista S. Doerfler, courtesy *Fine Homebuilding*, © The Taunton Press, Inc.; except p. 81 © Arch Wood Protection; p. 83 (bottom) © Woodsmart Solutions, Inc.; and p. 84 (bottom) © Trimax Building Products.

p. 86: LVLs: A Strong Backbone for Floor Framing by John Spier, issue 173. Photos by John Fournier, courtesy *Fine Homebuilding*, © The Taunton Press, Inc.; except pp. 88 and 91 by Daniel S. Morrison, courtesy *Fine Homebuilding*, © The Taunton Press, Inc.; and p. 94 by by Roe A. Osborn, courtesy *Fine Homebuilding*, © The Taunton Press, Inc.; Drawings by Chuck Lockhart, courtesy *Fine Homebuilding*, © The Taunton Press, Inc.

p. 96: Save Time with Factory-Framed Floors by Fernando Pagés Ruiz, issue 189. Photos by Rob Yagid, courtesy *Fine Homebuilding*, © The Taunton Press, Inc.; except pp. 96 and 97 by Jeff Beebe, courtesy *Fine Homebuilding*, © The Taunton Press, Inc.; and p. 97 (inset) by Fernando Pagés Ruiz, courtesy *Fine Homebuilding*, © The Taunton Press, Inc.; Drawings p. 97 courtesy Millard Lumber and p. 99 by Dan Thornton, courtesy *Fine Homebuilding*, © The Taunton Press, Inc.

p. 104: Mudsills: Where the Framing Meets the Foundation by Jim Anderson, issue 157. Photos by Ron Ruscio, courtesy *Fine Homebuilding*, © The Taunton Press, Inc.; Drawings by Dan Thornton, courtesy *Fine Homebuilding*, © The Taunton Press, Inc.

p. 112: The Well-Framed Floor by Jim Anderson, issue 160. Photos by Chris Green, courtesy *Fine Homebuilding*, © The Taunton Press, Inc.; Drawings by Chuck Lockhart, courtesy *Fine Homebuilding*, © The Taunton Press, Inc.

p. 120: Framing and Sheathing Floors by Rick Arnold and Mike Guertin, issue 117. Photos by Roe A. Osborn, courtesy *Fine Homebuilding*, © The Taunton Press, Inc.

p. 131: Built-Up Center Beams by Rick Arnold and Mike Guertin, issue 144. Photos by Roe A. Osborn, courtesy *Fine Homebuilding*, © The Taunton Press, Inc.; except

p. 135 by Tom O'Brien, courtesy *Fine Homebuilding*, © The Taunton Press, Inc.; Drawings by Dan Thornton, courtesy *Fine Homebuilding*, © The Taunton Press, Inc.

p. 140: Installing Floor Trusses by Brian Colbert, issue 114. Photos by Steve Culpepper, courtesy *Fine Homebuilding*, © The Taunton Press, Inc.; Drawings by Christopher Clapp, courtesy *Fine Homebuilding*, © The Taunton Press, Inc.

p. 148: Framing Floors with I-Joists by Rick Arnold and Mike Guertin, issue 108. Photos by Roe A. Osborn, courtesy *Fine Homebuilding*, © The Taunton Press, Inc.; except pp. 152 (top) and 154 (bottom left) by Rick Arnold and Mike Guertin, courtesy *Fine Homebuilding*, © The Taunton Press, Inc.

p. 157: Supporting a Cantilevered Bay by Mike Guertin, issue 136. Photos by Roe A. Osborn, courtesy *Fine Homebuilding*, © The Taunton Press, Inc.

p. 163: 6 Ways to Stiffen a Bouncy Floor by Mike Guertin and David Grandpré, issue 184.Photos by Mike Guertin, courtesy *Fine Homebuilding*, © The Taunton Press, Inc.; except p. p. 163 by Daniel S. Morrison, courtesy *Fine Homebuilding*, © The Taunton Press, Inc.; p. 167 by Justin Fink, courtesy *Fine Homebuilding*, © The Taunton Press, Inc.; and pp. 170 and 172 by Krista S. Doerfler, courtesy *Fine Homebuilding*,

© The Taunton Press, Inc.; Drawings by Don Mannes, courtesy *Fine Homebuilding*, © The Taunton Press, Inc.

p. 173: Careful Layout for Perfect Walls by John Spier, issue 156. Photos by Roe A. Osborn, courtesy *Fine Homebuilding*, © The Taunton Press, Inc.; Drawings by Vince Babak, courtesy *Fine Homebuilding*, © The Taunton Press, Inc.

p. 182: Setting the Stage for Wall Framing by Jim Anderson, issue 165. Photos by Chris Green, courtesy *Fine Homebuilding*, © The Taunton Press, Inc.; Drawings by Dan Thornton, courtesy *Fine Homebuilding*, © The Taunton Press, Inc.

p. 192: Laying Out and Detailing Wall Plates by Larry Haun, issue 126. Photos by Andy Engel, courtesy *Fine Homebuilding*, © The Taunton Press, Inc.; except p. 198 (top) by Larry Haun, courtesy *Fine Homebuilding*, © The Taunton Press, Inc.; Drawings by Rick Daskam, courtesy *Fine Homebuilding*, © The Taunton Press, Inc.

p. 202: Not -So-Rough Openings by John Spier, issue 176. Photos by Roe A. Osborn, courtesy *Fine Homebuilding*, © The Taunton Press, Inc.; Drawings by Dan Thornton, courtesy *Fine Homebuilding*, © The Taunton Press, Inc.

p. 208: Framing Curved Walls by Ryan Hawks, issue 148.

Photos by Tom O'Brien, courtesy *Fine Homebuilding*, © The Taunton Press, Inc.; Drawings by Dan Thornton, courtesy *Fine Homebuilding*, © The Taunton Press, Inc.

p. 216: Framing Big Gable Walls Safely and Efficiently by Lynn Hayward, issue 181. Photos by Daniel S. Morrison, courtesy *Fine Homebuilding*, © The Taunton Press, Inc.; Drawings by Heather Lambert, courtesy *Fine Homebuilding*, © The Taunton Press, Inc.

p. 226: Raising a Gable Wall by John Spier, issue 122. Photos by Roe A. Osborn, courtesy *Fine Homebuilding*, © The Taunton Press, Inc.

p. 235: Better Framing with Factory-Built Walls by Fernando Pagés Ruiz, issue 169. Photos by Fernando Pagés Ruiz, courtesy *Fine Homebuilding*, © The Taunton Press, Inc.

p. 241: Curved Ceiling? No Problem by Michael Chandler, issue 185. Photo by Seth Tice-Lewis, courtesy *Fine Homebuilding*, © The Taunton Press, Inc.; Drawings by Toby Welles, courtesy *Fine Homebuilding*, © The Taunton Press, Inc.

p. 244: Framing Cathedral Ceilings by Briab Saluk, issue 118. Photos by Greg Morley, courtesy *Fine Homebuilding*, © The Taunton Press, Inc.; except p. 245 by Scott Gibson, courtesy *Fine Homebuilding*, © The Taunton Press, Inc.; Models by Linden Frederick, courtesy *Fine Homebuilding*, © The Taunton Press, Inc.

p. 254: Ceiling Remodel: From Flat to Cathedral by Mike Guertin, issue 192. Photos by Charles Bickford, courtesy *Fine Homebuilding*, © The Taunton Press, Inc.; Drawings by Don Mannes, courtesy *Fine Homebuilding*, © The Taunton Press, Inc.

p. 264: Open Up the Ceiling with a Steel Sandwich by Michael Chandler, issue 178. Photos by Daniel S. Morrison, courtesy *Fine Homebuilding*, © The Taunton Press, Inc.; except p. 265 by Seth Tice-Lewis, courtesy *Fine Homebuilding*, © The Taunton Press, Inc.; Drawings by Heather Lambert, courtesy *Fine Homebuilding*, © The Taunton Press, Inc.

INDEX

Taunton's FOR PROS BY PROS Series
A Collection of the best articles from Fine Homebuilding magazine

OTHER BOOKS IN THE SERIES:

Taunton's For Pros By Pros:
RENOVATING A BATHROOM

ISBN 1-56158-584-X
Product # 070702
$17.95 U.S.
$25.95 Canada

Taunton's For Pros By Pros:
BUILIDING ADDITIONS

ISBN 1-56158-699-4
Product # 070779
$17.95 U.S.
$25.95 Canada

Taunton's For Pros By Pros:
BUILIDING STAIRS

ISBN 1-56158-653-6
Product # 070742
$17.95 U.S.
$25.95 Canada

Taunton's For Pros By Pros:
BUILT-INS AND STORAGE

ISBN 1-56158-700-1
Product # 070780
$17.95 U.S.
$25.95 Canada

Taunton's For Pros By Pros:
EXTERIOR SIDING,
TRIM & FINISHES

ISBN 1-56158-652-8
Product # 070741
$17.95 U.S.
$25.95 Canada

Taunton's For Pros By Pros:
FINISH CARPENTRY

ISBN 1-56158-536-X
Product # 070633
$17.95 U.S.
$25.95 Canada

Taunton's For Pros By Pros:
FOUNDATIONS AND
CONCRETE WORK

ISBN 1-56158-537-8
Product # 070635
$17.95 U.S.
$25.95 Canada

Taunton's For Pros By Pros:
RENOVATING A KITCHEN

ISBN 1-56158-540-8
Product # 070637
$17.95 U.S.
$25.95 Canada

Taunton's For Pros By Pros:
FRAMING ROOFS

ISBN 1-56158-538-6
Product # 070634
$17.95 U.S.
$25.95 Canada

Taunton's For Pros By Pros:
BUILDING PORCHES
AND DECKS

ISBN 1-56158-539-4
Product # 070636
$17.95 U.S.
$25.95 Canada

Taunton's For Pros By Pros:
ATTICS, DORMERS,
AND SKYLIGHTS

ISBN 1-56158-779-6
Product # 070834
$17.95 U.S.
$25.95 Canada

Taunton's For Pros By Pros:
FRAMING FLOORS,
WALLS AND CEILINGS

ISBN 1-56158-758-3
Product # 070821
$17.95 U.S.
$25.95 Canada

Taunton's For Pros By Pros:
TILING

ISBN 1-56158-788-5
Product # 070843
$17.95 U.S.
$25.95 Canada

Taunton's For Pros By Pros:
ROOFING, FLASHING
AND WATERPROOFING

ISBN 1-56158-778-5
Product # 070833
$17.95 U.S.
$25.95 Canada

Taunton's For Pros By Pros:
CABINETS AND
COUNTERTOPS

ISBN 1-56158-806-7
Product # 070858
$17.95 U.S.
$25.95 Canada

Taunton's For Pros By Pros:
WINDOWS AND DOORS

ISBN 1-56158-808-3
Product # 070860
$17.95 U.S.
$25.95 Canada

For more information contact our website at: www.taunton.com